国际工程科技发展战略高端论坛
International Top-level Forum on Engineering Science and Technology Development Strategy

农业航空技术

罗锡文　赵春江　编著

中山大學出版社
SUN YAT-SEN UNIVERSITY PRESS
·广州·

图书在版编目(CIP)数据

农业航空技术/罗锡文，赵春江编著．—广州：中山大学出版社，2021.8
ISBN 978-7-306-07149-1

Ⅰ．①农…　Ⅱ．①罗…　②赵…　Ⅲ．①农业—航空学—文集　Ⅳ．①S25-53

中国版本图书馆 CIP 数据核字(2021)第 039427 号

NONGYE HANGKONG JISHU

出 版 人：王天琪
策划编辑：张　蕊
责任编辑：张　蕊
封面设计：林绵华
责任校对：梁俏茹
责任技编：何雅涛
出版发行：中山大学出版社
电　　话：编辑部 020-84111997，84113349，84111996，84110779
　　　　　发行部 020-84111998，84111981，84111160
地　　址：广州市新港西路 135 号
邮　　编：510275　　　　传　　真：020-84036565
网　　址：http://www.zsup.com.cn　　E-mail：zdcbs@mail.sysu.edu.cn
印 刷 者：广州市友盛彩印有限公司
规　　格：787mm×1092mm　1/16　9 印张　200 千字
版次印次：2021 年 8 月第 1 版　2021 年 8 月第 1 次印刷
定　　价：50.00 元

内 容 提 要

农业航空是现代农业的重要组成部分，它的发展水平是农业现代化水平的重要标志之一。目前，世界农业航空产业发展迅速，但中国农业航空的应用水平还不高，不能满足现代农业发展的需求。2017 年，中国工程院在北京举办了“国际工程科技发展战略高端论坛——农业航空技术”，与会嘉宾分别来自中国、美国、加拿大、澳大利亚、德国、法国、西班牙、韩国、日本和比利时。这些顶尖专家围绕“加强国际合作交流，促进农业航空创新发展”这一主题进行了主题演讲和学术讨论。本书收集了这次论坛的学术论文和部分专家的报告，涵盖了农业航空模式、农业航空关键技术、农业航空服务体系、农业航空相关政策等领域的最新成果。

本书是中国工程院“国际工程科技发展战略高端论坛”系列丛书之一，所收文章具有重要的参考价值，可供从事农业航空技术的科研人员参阅。

中国工程院副院长田红旗院士致辞①

各位专家、各位来宾，女士们、先生们：

大家上午好！

很高兴有机会和大家共同交流我国农业航空事业的发展。首先，请允许我代表这次论坛的主办单位——中国工程院，对参加“国际工程科技发展战略高端论坛——农业航空技术”的嘉宾和代表表示热烈的欢迎！向科技部、农业农村部、有关行业学会/协会/企业、有关大学及科研机构对本次论坛的支持表示衷心感谢！对为组织本次论坛会议而付出了辛勤劳动的单位表示诚挚谢意！

近年来，中国农业现代化建设取得了巨大成绩，粮食产量连续三年超过 1.2 万亿千克，农田有效灌溉面积占比、农业科技进步贡献率和主要农作物耕种收综合机械化率分别达到 52%、56% 和 63%，现代农机装备和先进科学技术对中国农业发展起到了重要的作用。当前，我国农业发展的内外环境正在发生深刻的变化，农业资源与生态环境问题、城镇化带来的农村劳动力短缺、农业产业结构失衡与供给侧结构性改革成为我们面临的重大挑战，要求我们进一步加快科技创新，强化现代物质技术装备支撑，走绿色生态可持续发展之路，大幅度提高资源利用率、土地产出率和劳动生产率。

农业航空是采用航空飞行器完成农业特定作业任务的一个重要领域，具有作业效率高、覆盖面积广和应急能力强等特点，在植物保护、森林防火、遥感测绘、耕地监测等方面具有巨大的应用潜力。随着航空技术与产业的发展，农业航空成为世界关注的热点。目前，全球农林用飞机已经接近 3 万架，年作业面积 1 亿公顷，飞机作业面积占总耕地面积的 17%，已从机型设计、加工制造、关键部件、航空作业到运维服务等方面形成了巨大的产业链。美国是农业航空应用技术最成熟的国家之一，在政策、法律、技术、应用、服务等方面建立了完善的体系，值得我们学习借鉴。

中国农业航空产业起步较晚，农用航空应用规模比较小，与我国现代农业发展的需求不相适应。2014 年，中央一号文明确提出“加强农用航空建设”的要求。随着我国土地流转和规模化经营的快速发展，特别是应对治理突发性农业病虫害的需要，农业航空具有广阔的发展前景。近年来，中国农业航空快速发展，特别是小型农用无人机发展迅

① 致辞时间为 2017 年 4 月 22 日。田红旗院士于 2016 年 6 月至 2017 年 6 月任中国工程院副院长，2017 年 5 月至今任中南大学校长。

速，在农业生产中发挥了重要的作用，但也面临着政策保障不足和存在安全隐患等重大问题。

中国工程院作为国家高端智库，高度关注我国农业航空事业的发展。2014 年 4 月，罗锡文院士联合 30 位院士，从安全监管、标准规范、科技创新、政策保证四个方面，向国家提出了加快推进我国农业航空植保产业创新发展的建议。2016 年，栾恩杰院士联合 15 位院士和专家，向国家提出了积极应对“低慢小轻”飞行器威胁，健全国家安全保障体系的建议。这次高端论坛，我们邀请了国内外农业航空领域产业界、科技界、政府部门和用户代表，重点围绕农业航空应用模式、农业航空关键技术、农业航空服务体系、农业航空相关政策等方面进行研讨交流，共享农业航空发展成果，借鉴发达国家的成功经验，促进中国农业航空事业健康、有序与创新发展。

最后，预祝本次论坛取得圆满成功，再次感谢大家对论坛的支持。

谢谢大家！

目录

综述

“国际工程科技发展战略高端论坛——农业航空技术”于2017年4月22—24日在北京隆重召开，论坛由中国工程院主办，中国工程院农业学部、华南农业大学、北京市农林科学院、国际农业与生物系统工程委员会［International Commission of Agricultural and Biosystems Engineering—CIGR（Commission Internationale du Genie Rural）］精准农业航空工作委员会联合承办。中国工程院田红旗、汪懋华、康绍忠、尹伟伦、孙九林、刘经南、罗锡文、陈温福、南志标、陈学庚、唐华俊、吴孔明、宋宝安等13位院士，美国、加拿大、澳大利亚、德国、法国、西班牙、韩国、日本、比利时等9个国家的19位特邀专家，国内22所高校、23所研究院所、34家农业航空企业、10个政府机构、6个学会/协会/企业的代表共计300余人参加了此次论坛。

中国工程院院士罗锡文担任本次论坛主席并主持论坛开幕式，时任中国工程院副院长田红旗、农业农村部农机化司司长李伟国、国家农业航空产业技术创新战略联盟副理事长/中国农村技术开发中心主任贾敬敦、北京市农林科学院院长李成贵、时任华南农业大学副校长廖明出席开幕式并致辞，时任科技部农村科技司副司长蒋丹平应邀出席了开幕式。

田红旗院士在致辞中指出，中国工程院作为国家高端智库，高度关注我国农业航空事业的发展。2014年4月，罗锡文院士联合30位院士，从安全监管、标准规范、科技创新、政策保证四个方面，向国家提出了加快推进我国农业航空植保产业创新发展的建议。2016年，栾恩杰院士联合15位院士、专家，向国家提出了积极应对“低慢小轻”飞行器威胁，健全国家安全保障体系的建议。本次论坛围绕我国农业航空模式、关键技术、服务体系、相关政策等内容进行了深入研讨，旨在促进我国农业航空事业更好地发展。

农业农村部农机化司李伟国司长在致辞中指出，农业航空技术是高端农业机械化技术，在现代农业生产特别是航化作业、病虫害防治等方面发挥了越来越重要的作用。农业农村部编制的《全国农业机械化发展第十三个五年规划》明确提出，到2020年，“农用航空作业面积明显增长”，并重点在华北平原地区和长江中下游地区稳步发展农用航空，在东北地区积极发展农用航空，在黄土高原及西北地区扩大农用航空作业面积。农业农村部和财政部商定，今年（2017年）将扩大农机新产品补贴试点范围，允许在适宜的地区开展植保无人飞机补贴试点，并将与有关部门一起，加快研究制定相关管理制度。

本次论坛围绕“加强国际合作交流，促进农业航空创新发展”这一主题，邀请中国农业大学汪懋华院士、武汉大学刘经南院士、中国农业科学院唐华俊院士、中国农业科学院吴孔明院士、新疆农垦科学院陈学庚院士、华南农业大学兰玉彬教授、美国农业部农业科学研究署航空应用小组 Bradley K. Fritz 高级研究员、法国农业科学院 Frederic Baret 教授和美国农业部森林服务署施药风险评估组项目主管 Harold W. Thistle 博士等做了大会报告。澳大利亚昆士兰大学施药与安全中心的 Andrew Hewitt 教授、加拿大 Fred Ramirez 博士、日本创新策略推进计划（Strategic Innovation-drive Project，SIP）项目负责人 Noboru Noguchi 教授等9位外国专家，中国农业大学、浙江大学、国家农业智能装备技术中心、农业农村部南京农机化研究所、北大荒通用航空公司、山东瑞达有害生物防控公司、珠海羽人农业航空公司和深圳大疆公司等10家机构的代表做了专题报告交流。

在论坛的自由发言研讨环节，来自国内外农业航空领域产业界、科技界、政府部门的代表和用户代表进行了互动交流和深入研讨，对农业航空发展的现状、问题、对策等发表了自己的观点。论坛还在会议现场进行了农业航空领域最新技术成果和产品展示，组织参会代表到北京国际展览中心参观了第八届中国国际农业航空技术装备展览会。

最后，论坛主席罗锡文院士根据论坛交流讨论情况，从农业航空的管理、模式、标准、创新、应用、政策6个方面对论坛成果进行了总结。他指出，第一，要明确农业航空的政府主管部门，统筹管理全国农业航空产业发展，制定产业发展规划，建立市场准入制度和管理规范，实现政府对农业航空产业技术产品质量和作业安全的有效监管，确保农业航空产业健康有序发展；第二，要进一步探讨农业航空的发展模式，研究有人驾驶固定翼飞机/直升机、无人直升机/多旋翼机的适用范围和条件；第三，要进一步加强中国农业航空技术标准和规范的制定，包括农用航空飞行器产品设计/生产/制造的技术标准、产品质量检测标准与质量管理体系、农业飞行作业操作人员岗位培训与资质认证体系、农业航空作业质量、防治效果评价和环境风险评估标准规范等；第四，要进一步加强中国农业航空关键技术的协同创新研究，重点是电/油动机的发动机和电池、手控模式和 GNSS 控制模式的飞控系统、航空施药专用药剂/部件和航空遥感平台及专用传感器等；第五，要以新型经营主体为载体，加大农业航空技术在农林业植保、播种、授粉、施肥、防火和监测等方面的应用；第六，要制定促进农业航空应用发展的相关政策，包括开放农区空域、简化审批程序、出台产业发展金融贷款和财政补贴及税收优惠政策等。

本次论坛对农业航空发展战略、技术、政策、法律等方面进行了深入的研讨交流，对促进中国农业航空技术与产业的健康、有序与创新发展具有重要意义。

参会专家名单

外方专家

ANDREW HEWITT　Professor/Director, the Center of Pesticide Application and Safety, University of Queensland, Australia（澳大利亚昆士兰大学施药与安全中心教授/主任）

YANBO HUANG　Professor/Lead Scientist, Crop Production Systems Research Unit, Agricultural Research Services（ARS）, United States Department of Agriculture（USDA）, USA（美国农业部农业科学研究署作物生产系统研究组教授/首席科学家）

HAROLD W. THISTLE　Program Manager, USDA Forest Service, Pesticide Risk Assessments, USA（美国农业部森林服务署施药风险评估组项目主管）

TOMONARI FURUKAWA　Professor, Virginia Tech Department of Mechanical Engineering, USA（美国弗吉尼亚理工机械工程系教授）

NING WANG　Professor, Oklahoma State University, USA（美国俄荷拉马州立大学教授）

DAVID MULLA　Professor, Department of Soil, Water, and Climate, University of Minnesota, USA（美国明尼苏达大学土壤、水与气候系教授）

NAIQIAN ZHANG　Professor, Kansas State University, USA（美国堪萨斯州立大学教授）

MARC VANACHT　Professor, AG Business Consultants, USA（美国农业战略咨询公司教授）

BRADLEY K. FRITZ　Senior Researcher, Aerial Application Technology Group, United States Department of Agriculture（USDA）, USA（美国农业部农业科学研究署航空应用小组高级研究员）

VICTORIA GONZALEZ-DUGO　Researcher, Spanish National Research Council/Institute for Sustainable Agriculture, Spain（西班牙国家研究委员会可持续农业研究院研究员）

NOBORU NOGUCHI　Professor, Research Faculty of Agriculture, Hokkaido University; Program Director, Cross-Ministerial Strategic Innovation Promotion Program（SIP）, Bureau of Science, Technology and Innovation, Cabinet Office, Government of Japan, Japan（北海道大学农学部教授，日本内阁府科学、技术和创新局“创新策略推进计划”项目首席专家）

KYEONG-HWAN LEE　Professor, Department of Rural & Bio-systems Engineering, Chonnam National University, Korea（韩国全南大学农村与生物系统工程系教授）

SUN-OK CHUNG　Professor, Department of Bio-systems Machinery Engineering, College of Agriculture and Life Science, Chungnam National University, Korea（韩国忠南大学生物系统工程学院教授）

GEORG BARETH　Professor, Geography, GIS & RS University of Cologn, German（德国科隆大学地理信息系统与遥感系教授）

JENS BONGARTZ　Professor, Fraunhofer Institute for High-Frequency Physics and Radar-Technology（FHR）Application Center for Multimodal and Airborne Sensing（AMLS）, Germany（德国弗劳恩霍夫应用研究促进协会教授）

FREDERIC BARET　Doctor, French Academy of Agriculture and Forestry, France（法国农科院博士）

QINGHAN DONG　Researcher, Flemish Institute for Technological, Vito, Belgium（比利时佛莱芒技术研究所研究员）

FRED RAMIREZ　President, AG-NAV Inc., Canada（加拿大 AG-NAV 公司董事长）

中方专家

田红旗　中国工程院院士，中国工程院副院长

康绍忠　中国工程院院士，中国工程院农业学部副主任，中国农业大学教授

汪懋华　中国工程院院士，中国农业大学教授

刘经南　中国工程院院士，武汉大学教授

孙九林　中国科学院地理科学与资源研究所研究员

尹伟伦　中国工程院院士，北京林业大学教授

南志标　中国工程院院士，兰州大学教授

陈温福　中国工程院院士，沈阳农业大学教授

唐华俊　中国工程院院士，中国农业科学院研究员

吴孔明　中国工程院院士，中国农业科学院研究员

陈学庚　中国工程院院士，石河子大学教授

宋宝安　中国工程院院士，贵州大学教授

罗锡文　中国工程院院士，华南农业大学教授

李伟国　农业农村部农业机械化管理司原司长

蒋丹平　科技部农村科教司原副司长

贾敬敦　中国农村技术开发中心原主任

刘　敏　农业农村部农业机械化试验鉴定总站原站长

刘恒新　农业农村部农业机械化技术开发推广总站原站长

刘　宪　中国农业机械化协会会长

杨　林　中国农业机械化协会农业航空分会主任

李成贵　北京市农林科学院院长

廖　明　华南农业大学原副校长

（注：会议时间为 2017 年 4 月 22 日至 24 日）

主题报告

智慧农业发展需要农业航空技术与服务型制造产业的创新支撑①

汪懋华②

1 中国农业发展进入新阶段——实施《国民经济和社会发展第十三个五年规划纲要》，加快推进农业现代化

“十三五”（2016—2020）期间是我国全面建成小康社会的决胜阶段，要实现新型工业化、信息化、城镇化和农业现代化同步发展，“农业是短腿，农村是短板”。2016年5月20日，中共中央、国务院印发了《国家创新驱动发展战略纲要》，明确了国家科技创新事业发展的战略目标：到2020年，使我国进入创新型国家行列；到2030年，使我国跻身创新型国家前列；到2050年，使我国建成为世界科技创新强国。2016年5月30日，习近平总书记在全国科技创新大会上提出：“要在我国发展新的历史起点上，把科技创新摆在更加重要位置。”报告强调需要优化科技事业发展总体布局的5个方面：“要夯实科技基础，在重要科技领域跻身世界领先行列；要强化战略导向，破解创新发展科技难题；要加强科技供给，服务经济社会发展主战场；要深化改革创新，形成充满活力的科技管理和运行机制；要弘扬创新精神，培育符合创新发展要求的人才队伍。”

《国民经济和社会发展第十三个五年规划纲要》（以下简称《纲要》）第四篇《推进农业现代化》指出：农业是全面建成小康社会和实现现代化的基础，必须加快转变农业发展方式，着力构建现代农业产业体系、生产体系、经营体系，提高农业质量效益和竞争力，走产出高效、产品安全、资源节约、环境友好的农业现代化道路。这包括“增强农产品安全保障能力”“构建现代农业经营体系”“提高农业技术装备和信息化水平”“完善农业支持保护制度”等内容。具体内容为提高粮食生产能力保障水平，加快推动农业结构调整，推进农村第一、第二、第三产业融合发展，确保农产品质量安全，促进农业可持续发展，开展农业国际合作；强调发展适度规模经营，培育新型农业经营主体，健全农业社会化服务体系；加强农业与信息化技术的融合，发展智慧农业，推进

① 本文根据作者于2017年4月22日在北京国际工程科技发展战略高端论坛发表的学术报告整理成文。

② 汪懋华，中国农业大学信息与电气工程学院教授，中国工程院院士，国际欧亚科学院院士，中国农业工程学会、中国农业机械学会名誉理事长。

农业信息化建设，推动信息技术与农业生产管理、经营管理、市场流通、资源环境等融合，实施物联网应用和推动农业大数据应用，增强农业综合信息服务能力；鼓励互联网企业建立产销连接的农业服务平台，加强发展涉农电子商务；突出以保障主要农产品供给、促进农民增收、实现农业可持续发展为重点，完善强农、惠农、富农政策，提高农业支持保障效能；持续增加投入，完善农产品价格和收储制度；创新农村服务等内容。

《纲要》专栏六明确提出要实施高标准农田建设，现代种业，节水农业，农业机械化，智慧农业，农产品质量安全，新型农业经营主体培育，农村第一、第二、第三产业融合发展八项农业现代化重大工程。2016 年 8 月 8 日，国务院发布《“十三五”国家科技创新规划》（以下简称《规划》）。《规划》中关于现代农业科技创新的部分指出：中国农业现代化建设取得巨大成绩，综合生产能力迈向新台阶，但部分农产品供求结构性失衡、农业发展方式粗放、农业竞争力不强，农业大而不强、多而不优问题突出，必须更新观念，科学谋划发展新思路。“发展现代农业科技”一节中的第23 条提出，要发展智慧农业技术。同年 10 月 17 日，国务院发布的《全国农业现代化规划（2016—2020 年）》提出，要实施五项“创新强农重大工程”，其中第四项是“实施智慧农业引领工程”。

因此，我们要把握国家推进农业现代化的发展战略、目标，考虑如何推进农业航空技术应用研究和产业化，同我国在“十三五”期间实现农业现代化的发展战略目标更好地结合起来。

2017 年是实施供给侧结构性改革的深化之年。中央经济工作会议提出，要确保“三去一降一补”获实质进展，深入推进农业供给侧结构性改革是其中的重点。随后召开的中央农村工作会议提出，要坚持把推进农业供给侧结构性改革作为农业农村工作的主线，培育农业农村发展的新动能，提高农业综合效益和竞争力。推进农业供给侧结构性改革，要在确保国家粮食安全的基础上，紧紧围绕市场需求的变化，以增加农民收入、保障有效供给为主要目标，促进农业农村发展由过度依靠资源消耗、主要满足“量”的需求，向绿色生态可持续、更加注重满足“质”的需求转变。这些内容在 2017 年中央一号文件关于聚焦农业供给侧结构性改革的部分也有详细的论述。

2017 年 4 月 10 日，国务院发布了《国务院关于建立粮食生产功能区和重要农产品生产保护区的指导意见》（以下简称《意见》）。《意见》提出了“两区”建设，即粮食生产功能区（稻谷、小麦、玉米）和重要农产品生产保护区（大豆、棉花、油菜、糖料、橡胶等）建设，要求落实“藏粮于地、藏粮于技”战略：在 3 年内要划定粮食生产功能区 0.6 亿公顷，重要农产品生产保护区 0.16 亿公顷；“两区”耕地占现有耕地 51.2%，占基本农田面积 67.3%；做到“两区”地块全部建档立册、上图入库，实现重要农产品种植结构全国“一张图”。

为推动我国农业航空应用创新驱动发展，首先要研究我国农业发展的战略需求和发展态势，在顶层设计、技术路线图研究的基础上推动我国农业航空应用技术的发展，加快推动我国农业现代化。

2 农业航空应用技术发展趋势

农业航空应用技术与服务有两种模式：一是有人驾驶的通用航空农业服务，二是无人机航空服务。前者在发达国家已有数十年的研究与应用历史，有比较成熟的应用基础研究和产业技术创新发展成果；我国部分地区，通用航空农林业应用也具有良好的发展基础。后者近两年以来在我国农业中的应用有了快速发展，相关应用实践探索也在不断推进。无人机农业应用研究在发达国家引起了广泛关注。据悉，中国深圳大疆民用无人机产品占有世界约70%的应用市场份额，但总体上无人机在农业领域应用的技术研究和市场培育仍处于发展的初级阶段。

2.1 有人驾驶通用航空在中国农林业应用的发展

2015 年年底，中国通用航空机场已超过 300 个，企业 281 家，在册各类飞机 1874 架，年飞行量约 73 万小时。2016 年 5 月 17 日，国务院出台了《促进通用航空业发展的指导意见》，提出要大力促进我国通用航空事业发展；计划到 2020 年，建成 500 个以上通用航空机场，通用航空器超 5000 架，年飞行量超过 200 万小时，通用航空业经济规模超 1 万亿元，在全国建设 50 个综合产业示范区。通用航空活动除涵盖公共航空运输以外，还包括农、林、牧、渔业等产业的作业飞行，我国需要培育一批具有市场竞争力的民间通用航空企业。我国通用航空仍处于初期发展阶段和相对较低水平，总量尚不足美国的 10%，发展潜力巨大。

我国通用农业航空发展在局部地区已有近半个世纪的历程，如早期在环渤海和黄海地区的滨海灭蝗卵、森林防火、病虫防治、飞播造林和人工气象干预等。农业通用航空在黑龙江垦区的应用已有 40 多年历史（2014 年已拥有农用飞机 87 架），基础设施较为完善；拥有先进的指挥、通信导航和气象设备，以及一支农艺、航空技术骨干和后勤保障队伍。北大荒通用航空有限公司以其自身为主体，吸收其他通用航空公司组建农业航空作业联合体，共同完成垦区航化作业。垦区农业航空环境优越，地势平坦，耕地集中，种植连片，高压线路和防护林带区划科学、布局合理，非常适合农用航空的飞机作业。2016 年，北大荒公司已有 100 架作业机群，年作业能力近 4000 万亩①，年飞行时间 2.5 万小时，农业航空作业面积占总耕地面积 54% 以上。

2.2 国内无人机应用发展现状与问题

目前，我国农用无人机市场处于快速发展阶段，但在农业领域的应用技术尚不成熟。虽然发展很快，但订单零散、品种较多、市场混乱，还没有形成集约化推进的应用规模，在研发和市场运行方面存在很多亟待解决的技术和制度性问题。这都对农用无人

① 1 亩≈666.67 平方米。

机市场的推广有着明显的影响。

中国的民用和消费级无人机产业发展在国际上处于领先水平。目前，中国制造的该类无人机占全球市场近70%的出口份额。民用无人机也受到农业用户的青睐，特别是无人机的植保应用，可以快速获取作物长势空间分布信息，及时获取农业灾害信息。

民用无人机制造产业风起云涌。作为高科技服务型制造行业，该产业的产品同质化问题严重，核心技术创新被弱化，存在以下较为严重的问题。

2.2.1 缺乏对口监管部门

我国尽管在2010年出台了1000米以下低空空域将逐步对民用领域开放的政策，但对农业航空的管理仍沿用现有民用航空标准，民航法规体系中的相关条款对农业应用存在针对性不强、结合不紧密等问题，在国家层面、行业内以及行业间缺少支持农业航空应用的整体发展规划。近年来，我国小型无人机发展迅速，但安全隐患突出，对口主管部门不明确，管理条例存在诸多空白。2009年的《民用无人驾驶航空器系统空中交通管理办法》并未对无人机农业应用指出一条明确的指导路线，更多的是限制无人机的运用以避免其对民航的干扰。

2.2.2 缺乏系统的安全管理

无人机农业应用的安全性是需要严肃考虑的问题，市场现存的无人机由于外形和结构多样，我们难以对其进行系统的等级管理。虽然中国民用航空局对适航许可、培训机构认证和培训人员认证都有要求，但还没有引起厂家和用户的重视，市场上的无人机仍存在安全隐患。

农民抗风险能力低。发生事故后，如造成较大损失，甚至造成人员伤亡，往往无法明确责任，也没有相应的事故鉴定机构。

航空保险类产品不完善。多数无人机保险是应对坠毁之后产生的有价物品赔偿，理赔程序和定损方式不完善，缺少由于无人机出现故障导致的误伤、误工等突发意外保险等。

2.2.3 无人机产品质量参差不齐

目前，国内市场上常见的无人机种类大多由航模或者航摄产品改进或改装而成，生产者在强度、净载、动载测试、长时间作业稳定性、复杂农田环境适应性和使用寿命等方面缺乏仔细认真的考虑，缺乏工业级制造解决方案。无人机质量检验标准的制定势在必行。无人机标准包括了条例和认证、飞机测试、维修与护理、通信、飞机监管、操作规范、安全性能、整机、部件的设计要求等内容，美国、日本、国际粮农组织等都有相关标准出台，但我国还没有相关的无人机标准。

2.2.4 结构设计有待提高

很多植保无人机基于对载重的需求，盲目追求更多旋翼和更大升力的桨，更改了原有空气动力和重力布局，降低了飞行器的安全性，存在结构设计不合理的问题。未经验证的空气动力和重力布局使荷载系数无法提升，增加了能耗，却无法发挥飞机本身的能力，这也使得能源耗费大；载重低，大部分在 10 ～ 15 千克；操作烦琐，培训时间长。“独特”的设计导致结构偏于复杂，学员需要花费更长的时间去了解和适应新的结构，检修麻烦；也使得药液装箱、电池更换、燃料添加等工序无法简化，耗费大量时间。这些都降低了无人机的作业效率。无人机的稳定性也有待提升。由于在喷药过程中，药箱中的液体位是变化的，而无人机在进行转弯、急停等操作时，药箱中的液体会由于流动性而改变无人机的重心，因而液体惯性容易使无人机失稳。

2.2.5 缺乏专门适应航空喷药的设备

我国的航空喷雾一直使用常规喷雾设备，针对性差，雾滴飘移严重。当飞机拐弯或处于非靶标区时，“拖尾巴”现象屡见不鲜。农药的有效利用率低，有些机器 80% 的药液洒落或飘移进入非靶标环境中。这不仅降低了防治效果，增加了成本，而且还会危及非靶标区域的敏感动植物、人类健康和生存环境。作业流程细节粗糙的问题也需要引起重视：无人机行进的速度高于人工喷药行走的速度，却仍沿用人工喷药的喷嘴和流量；缺乏对喷药过程中雾滴沉降与药液飘散规律的研究，在喷洒设备、喷施部件、作物表面黏附效果评价等方面仍缺乏研究；雾滴飘散严重。航空施药是一个复杂的物理过程，药剂从药液箱向生物靶标的剂量传递过程中，要经过喷头雾化、空中飞行、靶标撞击、沉积分布、吸收传导等一系列过程。在这个过程中，雾滴飘失总是不可避免地随着喷雾作业发生，是影响雾滴到达预定目标、造成农药浪费和环境污染的主要因素。

2.2.6 无人机自动化程度低

目前，一台植保无人机一般需要两个或两个以上的无人机操作人员：一个操作无人机，一个观测并负责加药等工作。如何针对地面情况进行准确喷药、喷过的地方不再喷药、对农田空白的地方停止喷药、降低农药无效损耗、延长单次起飞药箱容积的利用持续时间等是目前亟须解决的问题。

无人机导航方式单一，难以适应中国多变的地形。目前，无人机导航方式多以全球定位系统（global positioning system，GPS）为主要导航方式，但在山区丘陵地区等地形变化较大、树荫遮蔽的地区，GPS 信号被严重削弱，而差分 GPS 基站布置价格昂贵。缺乏精确导航方式，直接制约植保喷药的自动化进程。无人机姿态增稳的调整与操控方式复杂，飞行稳定性难控制。在接近植物冠层的时候，会产生向上的紊流，所处气流环境复杂，无人机如何迅速反应、保持稳定、提高安全、是亟待解决的问题。

2.2.7 无人机类型与应用仍然较少

目前，开发出来的无人机农业应用领域有限，主要集中在植保喷药、育种授粉、农田信息采集等几个方面；应该扩大其在农、林、牧、渔业中的应用研究。

2.2.8 民用无人机造成的对国家和社会安全的隐患不容忽视

据美国巴德学院统计，2013 年年底至 2015 年 9 月，全球无人机或遥控飞机与民航飞机共发生 327 起危险接近事件，导致重大事故与伤亡的风险在加大。2015 年 4 月，日本首相官邸上方突现一架无人机，上面装有摄像头和发烟筒装置。同日，一架电视台摄像机用无人机坠落到英国驻日大使馆院内。2016 年 3 月 15 日，法国戴高乐机场险些出现无人机与法航客机“遭遇”的悲剧，幸亏副驾驶紧急启动人工驾驶模式才得以幸免。2016 年 3 月，英国空中飞行安全机构公布 2015 年 4 月至 10 月共有 23 起无人机与飞机擦身而过险些相撞的事件。近几个月来，媒体报道在我国杭州、昆明、贵阳等机场均发生过无人机威胁航空飞行安全的事件。2016 年 4 月 17 日下午，成都双流机场附近出现无人机，使 11 个航班被迫降落重庆机场。当前，以轻型无人机、航模为代表的轻飞行器发展迅速，在广泛应用的同时，因成本低廉、获取容易、探测识别难，极易被境内外各种敌对势力利用，成为暴力恐怖、文化渗透、颠覆分裂、军事侦察及打击的工具，对国家安全构成现实和潜在威胁。因此，我们必须加强无人机管控的法制保障、政策引领、技术支撑，促进其有序发展，为防范国家安全风险、健全国家安全体系发挥重要作用。根据中国航空器拥有者及驾驶员协会（Aircraft Owners and Pilots Association of China，AOPA-China）于 2015 年年底发布的《中国无人机报告》，我国无人机驾驶员仅有 1249 名，上万架无人机处于“黑飞状态”。根据中国民航局下发的《民用无人驾驶航空器系统驾驶员管理暂行规定》，AOPA-China 宣布已得到民航局飞行标准司批文，要以云计算为基础进行无人机监管。该监管范围覆盖 1500 米以下所有直升机、无人机等飞行器的低空监测，目前主要针对重量在 7 ～25 千克、150 米高度下作业的无人机，飞行时所有动作变化，包括航迹、高度、速度、位置、航向等数据都会被系统记录。

3 智慧农业发展与农业航空技术应用创新

近几年来，“智慧”一词被广泛应用于国民经济社会发展研究的众多领域，例如，智慧城市、智慧能源、智慧交通、智慧保健和智慧农业等。2008 年年末，为应对世界金融危机，IBM 公司向奥巴马总统提出建设智慧地球（Smart Planet）的概念：更透彻的感知和度量，更全面的互联互通，更深入的智慧化应用。三大信息科技动力推进了世界智慧进程，几乎任何系统都可以实现数字量化和互联，使人们能够做出智慧的判断和处理。

20 世纪 90 年代，计算平台与互联网应用的快速发展推动世界进入了由工业经济向信息经济发展的新时代。1993 年年末海湾战争结束后，美国宣布 GPS 民用化，发达国家农业科技界提出实践“精细农业”（precision agriculture，PA）的理念，同时也推动了信息化与农业装备现代化融合研究的创新发展。2010 年，联合国粮食及农业组织（Food and Agriculture Organization of the United Nations，FAO）发布了气候智慧农业（climate smart agriculture，CSA）新理念。它定义“气候智慧农业”是集成实现农业可持续发展的经济、社会和环境三维度的理解，以应对食物安全和气候变化的挑战，即“defining the CSA concept as：it integrates the three dimensions of sustainable development（economic，social and environmental）by jointly addressing food security and climate challenges”。

20 世纪 90 年代中期以来，精细农业在全球发达国家的广泛实践助推了智慧农业（smart agriculture，SA）创新发展。“精细农业”向“智慧农业”的助推发展经历了以下的过程。自 20 世纪 80 年代初起，提出实施定位土壤管理与定位作物生产管理（site-specific soil & crop management）；20 世纪 90 年代中起，提出实施精细农业（precision agriculture，PA 或 precision farming，PF）；21 世纪初，基于新一代信息与无线通信技术发展提出实施 SA 和 CSA 的实践。展望 2025 年前后，后智能农业（intelligent agriculture，IA）又会被推到农业系统创新发展的前沿。

自 1994 年起，我国农业工程科技界开始研究海外发达国家精细农业技术的早期实践，海外精细农业机械装备技术产品提供商开始探索中国巨大的市场需求潜力。与此同时，国内研发机构到发达国家开展积极的技术考察和交流。世纪之交，我国已经建立了一批精细农业技术研发机构和示范试验农场，开始了对国外相关先进技术的消化吸收与示范应用实践研究。

“精细农业”发展实践面对的挑战性问题主要包括：研究与开发（Research & Development，R & D）主要集中在技术创新推动思维的指导下开展，较少关注系统优化、成本节约和用户投入的经济效益分析研究；终端用户需要高新技术以解决农业生产经营提质增效所面对的问题。新一代信息与通信技术的普及应用应提供更为有效的农业提质增效、绿色发展、节本增收和广泛实践智慧农业经营的机会。

“智慧农业”是基于新一代信息与通信技术推进农业现代化发展的重要抓手。随着人口增长、资源制约和环境恶化，凡能有效提高农业土、水、肥、药、光、热等资源利用率和生态环境保护的技术，都将成为智慧农业发展研究的新热点。智慧农业的发展，需要加强不同学科间的交叉融合，创新研究方式，跨入信息科技应用研究发展的前沿。基于信息和智慧管理好复杂的农业系统，促进农业、农村信息化，包括数字化、网络化、泛在化、智慧化和智能化阶段性发展的过程。推进农业现代化的发展，要聚焦转变农业发展方式，提高土地产出率、资源利用率和劳动生产率。统筹城乡经济社会发展对为农业、农村服务的信息科技创新提出了迫切需求。在现阶段推动智慧农业创新发展，将引领提升为农业与农村服务的信息化科技创新能力，推进“信息化与农业现代化的深度融合发展”。

4 结束语

新一代信息技术正迅速融入各种应用领域，成为交叉汇聚学科的纽带。发展智慧农业创新研究，将引领新一代信息技术与农业农村现代化深度融合发展和智慧决策的现代农业经营管理和产业技术创新，加快转变传统农业生产和经营、管理、服务模式，为保障食物安全、生态安全、产业增效、农业增收，以及发展为农业和美丽乡村建设服务的智慧产业做出重大的贡献！

发展智慧农业，要强化问题意识，坚持问题导向的创新驱动发展研究，要多深入农村调查研究，多了解不同地区农业农村不平衡、不充分发展的问题。推动智慧农业的创新实践，要适应不同地区农业发展所面临的问题，注重不同地区农业、农村与农民的发展机遇与潜力，让新一代信息技术与农业现代化深度融合发展研究，走上适应农业、农村自身发展的新道路。

作者简介

汪懋华，中国农业大学信息与电气工程学院教授，中国工程院、国际欧亚科学院院士。农业工程学科——现代农业装备、信息与电气工程科技专家，当代我国农业工程学科与教育事业承前启后的开拓者之一。1956 年毕业于原北京农业机械化学院农业机械系。1962 年夏于莫斯科国立农业大学获电气自动化技术科学副博士学位回国后到母校任教。1984 年至 1990 年，任原北京农业工程大学副校长。1991 年年初，经原国家教委选派，到泰国曼谷亚洲理工学院（国际性研究生院）担任客座教授两年。1986 年至 2003 年，先后兼任国务院学位委员会“农经、农业机械化”“农业工程”学科评议组成员兼召集人。先后曾任农业农村部科技委委员、常务委员；中国农业工程学会副理事长、理事长、荣誉理事长；中国农业机械学会副理事长、名誉理事长；全国高等农业院校教学指导委员会委员、副主任委员兼农业工程学科组长；曾受聘为联合国粮农组织农业工程专家组成员、美国堪萨斯州立大学客座教授、国际农业工程协会（CIGR）农业电气化与能源理事会副理事长。2006 年，被授予 CIGR 会士称号。2010 年，获 CIGR 杰出贡献奖。作为英国农业工程师学会会士，曾兼任电子技术与计算机应用专家组成员、英国《农业工程文摘》《食品控制》、荷兰《农业电子学与计算机》、欧洲《农业生物系统工程》等期刊的国际编委会委员；现任中国农业大学“教育部现代精细农业系统集成研究重点实验室”“农业农村部农业信息技术学科群”和设施农业学科群学术委员会主任等。20 世纪 90 年代中期，汪懋华院士根据国际农业工程的发展趋势，在中国率先传播“精细农业”的概念和知识，科学解析发达国家迅速发展中的相关工程科技与系统集成技术的研究进展。近 10 多年来，他紧跟时代步伐，在国内积极倡导和引领农业物联网技术及“智慧农业”系统创新发展研究，致力于农业信息感知、农业与农村信息化、新一代信息与通信技术的应用及科研成果产业化研究。

对促进我国农业航空技术健康发展的思考

罗锡文①

1 引言

农业航空是指采用低空载人或无人飞行器进行农业、林业、牧业、渔业生产和抢险救灾作业的飞行。[1]农业航空技术具有作业质量好、作业效率高、作业成本低以及环境适应性强等突出优点，是世界公认的先进农业生产技术，是农业生产现代化的重要标志之一。[2]尤其是农用无人机作业时能产生下旋气流吹动叶片，有利于雾滴沉积于目标作物叶片正反面[3]，且不受作物长势限制，能在作物生长中后期作业，对作物损伤小。[4]

农业航空起源于20世纪初。1918年，美国采用飞机喷洒砷素剂防治牧草害虫，随后，农业航空技术在加拿大、苏联、德国和新西兰等国家得到了应用与发展。目前，农业航空技术比较成熟的国家有美国、日本、俄罗斯、澳大利亚、加拿大和巴西等。[5]我国农业航空发展起源于20世纪50年代。1951年5月22日至23日，我国采用C－46型飞机在广州市上空连续2天执行了41架次的灭蝇任务。进入21世纪以来，得益于新型飞行器和电子信息技术的突破，世界农业航空科技创新及其产业发展取得了新的成就。

近年来，我国农业航空技术，特别是无人机技术发展非常迅速，在很多领域得到了推广应用，但与农业航空发展比较成熟的国家相比，我国的农业航空尚处于起步阶段，主要存在以下问题：①管理制度尚不完善，没有明确的管理部门和管理政策；②没有明确的发展模式；③没有统一的行业标准；④没有专业的创新团队；⑤缺乏新型的农用飞机使用载体；⑥在很多领域应用技术还不成熟；⑦缺乏完善的扶持政策。为了促进我国农业航空技术健康、有序地发展，本文分析了我国农业航空技术在管理、模式、标准、创新、载体、应用和政策7个方面的发展现状，指出了存在的问题，提出了解决问题的思路，旨在为促进我国农业航空技术的健康发展提供指导。

2 管理

无人机技术是农业航空技术的重要组成部分，因起降方便、操作灵活、购置成本低

---

① 罗锡文，华南农业大学南方农业机械与装备关键技术教育部重点实验室主任，华南农业大学工程学院教授。

而受到广泛青睐。[6]但无人机购买和使用门槛低，机具质量、操作人员的技术水平和作业质量参差不齐。自无人机投入市场以来，各种事故时有报道。我国无人机的管理制度尚不完善，亟待加强。

欧洲国家意识到农业航空技术，尤其是无人机技术的应用将会给世界经济、科技和工业带来巨大的影响，因此，较早制定了相关管理政策，以提高民众意识。[7]欧盟国家的无人机管理制度由欧洲联合航空局（European Joint Aviation Authorities，EJAA）和欧洲航空航行安全组织（European Organization for Safety of Air Navigation，EOSA）于2002年联合制定，规定无人机的操作必须在欧洲空中交通管理系统（European Air Traffic Management System，EATMS）的管制范围内。[8]2002年，英国发布了民用和军用无人机管理政策CAP 722，包括无人机的适航性、设计、生产、维修和操作等[9]，欧洲联合航空当局将英国这项政策作为欧洲轻型无人机系统的基本管理标准。[10]美国的无人机技术基于军用航空技术，1936年成立的航空模型学会（Academy of Model Aeronautics，AMA）为无人机管理制度的制定做出了重要贡献。联合航空管理局（Federal Aviation Admi-nistration，FAA）在FAA《2012年现代化与改革法案》会议中提出在2015年将无人机管理归入美国国家空域管理系统（National Airspace System，NAS），其中安全性是重点关注的问题。[11]澳大利亚的无人机以民航安全管理局（Civil Aviation Safety Authority，CASA）颁布的《民用航空安全条例》（*Civil Aviation Safety Regulations*，*CASR*）为管理依据。在澳大利亚申请使用无人机需要提供无人机相关信息并严格按民航安全管理局申请程序执行，使用超过25千克的无人机须通过无人机操作培训、注册申请并取得执照。[12]几个国家和地区的无人机管理实施政策及组织机构如表1所示。

表1　不同国家和地区无人机管理实施政策及组织机构

国家或地区	无人机管理组织机构	政策及内容
欧盟	European Joint Aviation Authorities（EJAA），European Organization for Safety of Air Navigation（EOSA）	安全性、保密性、适航性、操作批准、维修、执照
美国	Academy of Model Aeronautics（AMA），Federal Aviation Administration（FAA）	National Airspace System（NAS） 操作、认证要求、安全和保密性
英国	Civil Aviation Authority（CAA）	CAP 722 安全性、适航性，操作职责、不同质量飞行器的最高飞行高度和速度
澳大利亚	Civil Aviation Safety Authority（CASA）	*Civil Aviation Safety Regulations*（*CASR*） 飞行器质量、风险性、适航性操作、法律责任风险、执照

明确的管理部门和完善的管理制度是以上国家和地区的无人机产业健康发展的重要前提，相关政策和规章制度规范了无人机的使用，事故发生时有问责部门，减少了事故的发生，促进了无人机产业的健康发展。目前，我国当务之急是从以下四个方面规范农业航空的管理：①明确农业航空产业的政府主管部门，制定农业航空飞行器产品设计、生产、制造的技术标准，建立产品质量检测标准与质量管理体系；②制定产业发展规划，统筹管理全国无人机产业发展；③建立市场准入制度和管理规范；④实现政府对无人机产品质量、作业安全和作业质量的有效监管。

3 模式

农业航空作业机型大致可分为：有人驾驶固定翼飞机、有人驾驶直升机、无人驾驶直升机和无人驾驶旋翼飞机等（见图 1）。不同机型的结构不同，载荷不同，对起飞和降落条件的要求不同，作业时的飞行高度和速度不同，由机翼产生的风场也不同，适应于不同的作业场合和作业对象。[13]有人驾驶飞机载荷大、作业效率高，应用于连片大面积的农田作业，但因作业高度较高，导致雾滴飘移难以控制，易脱离目标区域，且对起降场地、使用地点和时间要求较高，使用的专用航空燃油成本高，加油不方便。有人驾驶作业方式的超低空飞行高度一般为 3 ～ 20 米，受低空气象条件（能见度、低空风切变等）影响大，存在安全隐患。[14]据统计，由超低空飞行造成的安全事故占民航安全事故的 80%。[15]

（a）有人驾驶固定翼飞机

（b）有人驾驶直升机

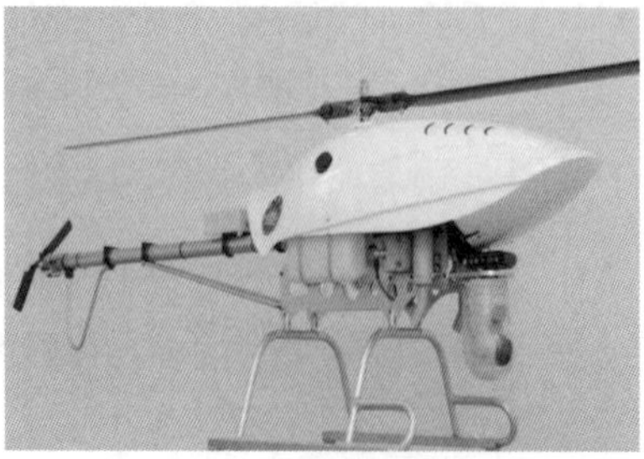

（c）无人驾驶直升机

（d）无人驾驶旋翼飞机

图 1　四种主要农业航空作业机型

微小型农用无人机相比有人驾驶飞机作业高度低，可以通过控制雾滴飘移减少对环境的污染，与全球导航卫星系统（Global Navigation Satellite System，GNSS）配合可实现较高精度的定位；旋翼产生的向下气流有利于提高雾滴在作物表面的附着性和在作物冠层的穿透能力，提高农药的有效利用率和防治效果；受周围障碍物的影响小；起降方便、作业成本低、灵活性强。[16, 17]但微小型无人机载荷小，续航时间短，作业效率低。

发达国家根据本国的情况选择适应其农业发展的航空技术发展模式。美国农业生产规模大，美国是世界上农业航空最发达的国家之一，农用飞机大约有20多种机型。由于美国幅员辽阔且地势平坦，因而美国的大部分机型是有人驾驶固定翼飞机。日本是一个人口多、耕地少的国家，每户平均耕地面积较小，种植规模较小，施药装备以小型机动喷雾机和无人直升机航空喷雾设备为主。

我国地形较复杂，既有幅员辽阔的北方平原地区，又有地形复杂的南方丘陵地区，对无人机的选型提出了更高的要求。目前，我国农业航空市场几乎包含了所有作业机型，但哪些机型更适应于我国农业的发展，是一个值得深入研究的问题。有人驾驶飞机与无人机应形成优势互补，对此，我们因地选型，各尽其用，提高我国航空作业的适应性。例如，我国北方地区地域辽阔，以平坦的大型农场为主，适宜采用载荷大、作业效率高的有人驾驶固定翼飞机或有人驾驶直升机；南方地区以丘陵等复杂地形的小地块为主，且电线杆等障碍物较多，适宜采用载荷小、对起降场地要求低的无人驾驶直升机或旋翼飞机。载重量为5千克的微小型无人机主要用于遥感航拍，载重10千克的无人机目前是航空施药的基本机型，载荷30千克的无人机具有较高的施药效率，载荷60千克的无人机作业效率更高，需要根据各地地形、地块大小和作物类型选择适宜的机型。

4 标准

根据《中华人民共和国标准化法》的规定：由我国各主管部、委（局）批准发布，在该部门范围内统一使用的标准，称为行业标准。机械、电子、建筑、化工、冶金、轻工、纺织、交通、能源、农业、林业和水利等部门，都有行业标准。制定行业标准的目的是规范该行业的各种行为，引导其发展方向。欧洲无人机联盟委员会（European Commission UAS Panel）指出，制定标准可促进欧洲无人机市场在欧洲以及国际发展的和谐性。[18]

很多发达国家对农业航空都制定了相关标准，行业内的组织和个人都必须严格按照标准规范其相关行为。欧盟国家对无人机技术（包括外观检测、负载大小测定和微型化功能性设备评估）、平台（包括管道和电力线的专用基础设施监控）和服务（如高海拔与地球同步旋转的带扩展功能的无人机卫星定位系统服务活动）制定了统一标准，任何使用无人机的组织和个人都必须遵守该组织制定的相关规定。欧洲航空安全局（European Aviation Safety Agency，EASA）规定无人机重量不能超过150千克。英国的轻型无人机政策包含无人机安全规则，大型模型协会（Large Model Association，LMA）可授权无人机生产和销售组织。[19] 2015 年，美国制定了集成开发无人机技术标准以及无人机质量认证标准等。澳大利亚对飞行员资格、无人机使用风险性、适航操作审批流程和法律责任风险性的评估与规定都设定了相关标准。

目前，我国的农业航空发展处于百家争鸣的状态。飞行器设计、制造，质量检测，飞行器操作岗前培训，飞行器操作执照发放，航空作业效果评估和购置补贴机型等都还没有统一的标准。中国有一句古话："没有规矩，不成方圆。"我国农业航空正处于起步阶段，标准的制定对其发展至关重要，为加快我国农业航空标准的制定，建议从以下五个方面入手：①农用航空飞行器产品设计、制造的技术标准；②农用航空飞行器产品质量检测标准与质量评估标准；③农业航空飞行作业流程规范、农业飞行作业操作人员岗位培训与资质认证标准；④农业航空作业质量、防治效果评价和环境风险性评估标准；⑤农业航空产品的质量监管与鉴定部门。标准出台后应坚决执行，不符合标准的农业航空飞行器或农业航空飞行器使用组织及个人不允许进入市场，以促进农业航空产业的健康发展。

5 创新

创新是一个产业发展的灵魂和驱动力。农业航空在农业生产中的应用主要包括播种、施肥、施药、遥感、测绘等，与农业航空相关的技术包括全球导航卫星系统、地理信息系统（Geographic Information Systems，GIS）、变量喷施系统、雾化系统、养分监测系统、土壤测绘系统、航空拍摄系统和航空施药模型等（见图 2）。

（a）航空播种

（b）航空施药

（c）航空遥感

（d）航空测绘

图 2　农业航空主要应用技术

国外农业航空起步较早，相关技术的创新和应用得到了很好的发展。航空施药是航空技术在农业生产应用中的一个重要方面，目前，国内外主要致力于航空精准施药。[20, 21]基于GPS研发了适用于无人机植保作业的精准施药系统，可对无人机作业时的飞行高度与飞行速度等参数进行精确控制，实现了病虫害定位管理与防治相结合的精准施药。国外比较成熟的航空变量施药系统有加拿大AG-NAV公司的AG-FLOW航空导航控制系统，可实时显示喷施区域状况信息、作业路线以及施药量等。Adapco公司生产的Wingman GX系统可指导并记录飞行器飞行路线，控制喷洒流量，通过气象传感器接收、处理气象信息，通过气象信息调整作业，可最大限度地减少农药漂移，优化喷洒质量。[22]国外基于遥感图像信息的农田土壤和养分等监测系统发展较快。美国农业部大面积病虫害管理研究中心航空应用技术团队采用MS4100多光谱相机获取棉花的航空遥感图像，生成棉花长势信息处方图，根据处方图进行精准施药。[23]日本研发了基于紫外线照相机的无人直升机遥感信息获取系统。该系统可从200～300米的低空全方位地获取水稻冠层高精度图像信息，将获取的紫外线图像经过蛋白质含量分类处理后得到的稻田生长信息提供给当地农协组织、农民或发布在因特网上，用于指导施肥作业。[24]

我国农业航空事业虽起步较晚，但近年来，在航空遥感、航空植保施药以及航空测绘等方面都取得了很大的进展。在快速发展的同时，建议从以下方面加强协同创新：①发动机。包括发动机（油）、电动机和电池的研发，建议成立专门的发动机和电池研究机构，研发质量精良的发动机和电池。②飞控系统。无人机的飞控系统包括手控和GNSS控制两种模式，手控模式由于人工操作造成的误差较大，GNSS精准控制系统是无人机控制模式的发展方向。③航空喷施技术。应加强喷施部件和喷施药剂的研究，如静电喷雾、防漂移等，适应于不同作业要求的喷施、防蒸发助剂的研究。④航空播种。应加强排种器以及精准定位系统等配套技术的研究，实现精确播种。⑤航空遥感和测绘。应重点加强飞行高度、续航能力、姿态控制、全天候作业能力以及大范围的动态监测能力等方面的研究。⑥农机农艺融合。农业航空各应用领域都应加强农机农艺融合，加强农机专家和农艺专家的交流与合作，发挥各自的优势，共同促进农业航空事业的发展。

6　载体

新型农业经营主体是指经营规模大，物质装备条件好，经营管理能力强，劳动生产率、资源利用率和土地产出率高，以商品化生产为主要目标的农业经营组织，包括专业大户、家庭农场、农民专业合作社、农业企业和（经营性）农业服务组织。农业机械社会化服务是指农业机械服务组织为其他农业生产者提供的耕整、播种、收获、排灌、植保以及相关的农业机械作业、供应、中介、租赁等有偿服务。我国户均耕地仅7～8亩，且一户多田块的情况比较普遍。我国农业人口人均耕地为0.13公顷，大约是美国的1/200、阿根廷的1/50、巴西的1/15、印度的1/2。尤其是南方地区耕地严重细碎化，对于一些大型昂贵的农业机械和公共农用设施，由于单块/片耕地无力承担或无须独立购置，提倡采用农业机械社会化服务模式。

发达国家的农业经营主体可分为企业化经营的农场、兼业农户（副业性农户）和合作社（合作经济组织）三种类型。美国的农业合作社与农业机械设施合作社已有200年历史，农业机械设施合作社为社员提供服务，同时收取手续费用于设施的购买、更新和维护，提供农药喷施、灌溉、施肥、病虫害防治等农业机械作业服务。农业机械化服务组织推进了农业机械化的发展，创造了农业现代化发展环境。从20世纪80年代开始，韩国大力倡导农民成立购买机械设备的联合体——营农团，营农团以5户或10户为单位，联合购买和使用农业机械。政府对营农团购买农业机械进行补贴，优惠率高达50%，且在此基础上，营农团可再申请40%的贷款。现在，韩国平均每两个村至少有1个营农团。

韩国与我国国情类似，以家庭小规模农业经营模式为主，其农业机械应用载体形式对我国农业机械化发展有很大的借鉴作用。农业机械社会化服务是发展农业机械化的重要载体。2000年以来，我国农业机械社会化服务组织发展迅速。到2013年年底，拥有农业机械原值50万元以上的农业机械大户和农业机械服务组织8.8万个，比2010年增加了4倍多。[25]截至2015年2月，全国农机作业服务专业户和农机合作社等各类服务组织数量分别超过530万个和170万个，每年完成作业服务面积近40亿亩，占全国农机作业总面积的2/3左右，农机田间作业服务收入超过2100亿元。[26]实践证明，农业机械社会化服务组织是促进农业增产和农民增收的重要组织之一。大力推进农业机械社会化服务，是构建“集约化、组织化、专业化、社会化”相结合的新型农业经营体系的重要措施，是解决农业生产“谁来种、种什么、怎么种”重大问题的现实途径，是实现农业机械化“全程、全面、高质、高效”发展的必然要求，也是促进我国农业航空发展的重要载体。

我国户均耕地面积小，每家每户都购买无人机不现实，也没有必要。建议以新型农业经营主体作为载体购买农用飞机，为农民提供农业航空作业服务。新疆的昌吉回族自治州供销合作社成立了一个农业航空作业服务组织，该合作社与农用飞机企业合作，购买农用飞机，为使用者提供农业航空作业服务。这种模式大大降低了农用飞机使用者的作业成本，提高了农用飞机利用率，有效促进了农业航空事业的发展。

7 应用

研究的最终目的在于应用。农业航空技术最初的应用始于病虫草害防治、播种和施肥。1918年，美国开始采用农用飞机进行航空施药防治病虫害。1929年，美国加利福尼亚州采用伊格尔罗克双翼飞机进行水稻撒播，随后航空播种水稻在美国迅速发展。目前，美国至少有90%的水稻采用航空播种。美国、日本、塞内加尔共和国等国家广泛使用飞机进行稻田肥料撒施，美国几乎全部稻田使用飞机施肥，美国、日本等国家在棉花播种和飞播造林等方面也取得了一定进展。[27]

目前，我国农业航空应用已经拓展到了授粉、农情信息采集等领域。[28]利用微小型无人机作业时产生的旋翼风场，进行了杂交稻制种辅助授粉，使原来 1 行或 2 行父本、8 行母本增加到了 8 行父本、60 行母本，扩展了父本与母本之间的宽度，提高了作业效率和结实率。低空遥感水稻信息获取平台采用微小型农用无人机配备多光谱相机、热成像仪等设备，用于水稻病虫害、氮素营养信息及长势信息的快速获取，可为大面积的水稻精准施肥提供决策依据，减少水稻生产中的肥料浪费。

农业航空在我国应用前景广阔。为促进我国农业航空产业又好又快地发展，当前必须重点解决以下关键问题：①安全性问题。任何技术的应用都应将安全问题放在首要位置。农业航空技术安全问题包括作业飞行器本身的安全问题和作业过程中环境安全的问题。②质量问题。质量问题包括农用飞机的制造质量和作业质量，农用飞机生产厂商必须保证农用飞机产品及配件的质量，农业航空作业实施者必须保证作业质量。③效益问题。相对于地面机械，农用飞机购置成本较高，对操作人员技术水平要求更高，培养操作人员所需要的费用和成本更高。因此，农业航空作业应努力提高经济效益、社会效益和生态环境效益等。④应用领域问题。目前，我国农业航空应用领域涉及植保、播种、施肥和农情信息采集、授粉、遥感、测绘等，建议进一步拓展我国农业航空应用领域，充分发挥农业航空作业高质量和高效率的优势。

8 政策

我国随着农业现代化步伐的加快，对农业航空作业的需求日益增加。政府的管理与扶持政策是确保农业航空健康快速发展的基础，农业航空的发展需要政府出台相关政策来推动，必须尽早明确农业航空管理机构，出台相关政策，对我国农业航空产业加强管理和规范。

很多农业航空产业发展较好的国家都有明确的管理机构，并出台了一系列政府扶持政策。为降低农业航空的作业成本，美国国会实行免除农用飞机每次起降 100 美元的机场使用费，从 2014 年起投入 73 亿美元支持该项政策。在全国农用航空协会（National Agricultural Aviation Association，NAAA）的推动下，自 2002 年以来，美国已投入约 700 万美元用于农业航空技术研发，参议院已通过议案将大力支持开发作业效率更高、使用成本更低的农业航空技术。俄罗斯农用飞机数量高达 1.1 万架，机型以有人驾驶固定翼飞机为主，年处理耕地面积约占总耕地面积 35% 以上。根据日本农民户均耕地面积小、以山地地形为主的特点，日本政府重点扶持发展农业航空直升机和微小型农用无人机。日本农用无人机航空协会（Japan Unmanned Aerial Vehicle Association，JUAV）目前共有 11 个单位会员，截至 2010 年 10 月，登记在册的微小型农用无人机有 2346 架，无人飞机操控手 14163 人，防治面积 96.3 万公顷，占航空作业的 38%。加拿大共有 169 个农业航空协会（Canada Agricultural Aviation Association，CAAA）会员。在国家政策的扶持下，包括农业航空在内的通用航空在巴西发展迅速。目前，巴西农业航空协会（Brazilian National Agricultural Aviation Association）共有143个单位会员。截至2008年3月，巴西农用飞机注册数量约1050架。

随着我国对粮食安全、生态安全、绿色植保等领域的发展需求的提高，国家对农业航空发展予以大力扶持。科技部和农业农村部在“十二五”科研规划中将农业航空应用作为重要支持方向。2014年中央一号文件《关于全面深化农村改革加快推进农业现代化的若干意见》第二条“强化农业支持保护制度”提出“加强农用航空建设”。

为促进我国无人机及农业航空事业健康发展，亟须出台一系列有利于农业航空发展的政府扶持政策。当前，建议先行出台下述政策：①简化审批程序，实行备案制度。②明确无人机开放的领空范围，并将无人机飞行高度严格控制在这个范围内，对超出领空范围的行为实行严格管制。③支持我国农用飞机和航空植保装备制造产业发展。④对农用飞机和航空植保装备的购置给予财政补贴。湖南、河南等省都出台了相关补贴标准，但是中央尚未出台正式的文件，建议政府制定相应的补贴标准。⑤对专业化和社会化的农业航空服务组织，在培训服务、融资贷款、经营税收等方面提供优惠政策。

9 结束语

农业航空技术具有作业质量好、作业效率高、作业成本低、对作物损伤小以及环境适应能力强等突出优点，是世界公认的先进农业生产技术。近年来，我国农业航空技术，尤其是无人机技术的发展非常迅速，在很多领域得到了推广应用，但与发达国家相比仍有较大差距。为促进我国农业航空技术健康发展，本文从管理、模式、标准、创新、载体、应用和政策七个方面提出了促进我国农业航空发展的建议，包括：①明确农业航空的主管部门，制定产业发展规划和管理规范，加强政府对农业航空的有效监管；②研究我国农用航空的发展模式，根据各地地形、地块大小和作物类型选择适宜的模式；③制定农业航空器设计制造标准、农业航空作业质量和环境适宜性标准，农业飞行操作人员岗位培训和资质认证标准；④加强农业航空发动机、电动机、电池、飞控系统和应用技术的创新研究；⑤将农业大户、家庭农场、专业合作社和农业企事业等新型主体作为农业航空技术应用的主要载体；⑥重视农业航空技术安全，提高质量，提高数量，拓展应用领域；⑦制定扶持农业航空发展的相关政策。

参考文献

［1］中国民用航空总局．通用航空经营许可管理规定［S］．中国民用航空总局令第176号．2007.

［2］BAE Y, KOO Y M. Flight attitudes and spray patterns of a roll-balanced agricultural unmanned helicopter［J］. Applied engineering in agriculture, 2013, 29 (5): 675-682.

［3］HUANG Y, HOFFMANN W C, LAN Y, et al. Development of a spray system for an unmanned aerial vehicle platform [J]. Applied engineering in agriculture, 2009, 25 (6): 803 - 809.

[4] 周志艳，臧英，罗锡文，等. 中国农业航空植保产业技术创新发展战略 [J]. 农业工程学报，2013，29 (24)：1 - 10.

[5] 薛新宇，梁建，傅锡敏. 我国航空植保技术的发展前景 [J]. 农业技术与装备，2010，(5)：27 - 28.

[6] 同 [3]。

[7] EUROPEAN DEFENCE AGENCY. European Commission UAS Panel: 5th workshop on research and development, discussion paper [S/OL]. [2012 - 02 - 09]. http://www.eda.europa.eu/docs/documents/UAS_RD_workshop_DiscussionPaper_final_1.pdf.

[8] JOINT AVIATION AUTHORITIES. UAV task-force final report: a concept for European regulations for civil unmanned aerial vehicles (UAVs) [S/OL]. [2004 - 05 - 11]. http://www.easa.europa.eu/rulemakingdocsnpaNPA_16_2005_Appendix.pdf.

[9] CIVIL AVIATION AUTHORITY. CAP 722: unmanned aerial vehicles operations in UK airspace-guidance [S/OL]. [2012 - 08 - 10]. http://www.caa.co.ukdocs33/CAP722.pdf.

[10] UAV task-force final report: a concept for European regulations for civil Unmanned Aerial Vehicles (UAVs) [S].

[11] FAA Modernization and Reform Act of 2012 (FMRA) [S]. Pub L No 112 - 95, 126 Stat 11 (Feb 2012).

[12] CASA. CASR Part 101, unmanned aircraft and rocket operations (2001) [S/OL]. http://www.casa.gov.au/scripts/nc.dll? WCMS:PWA:pc = PARTS101.

[13] 曾丽兰，王道波，郭才根，等. 无人驾驶直升机飞行控制技术综述 [J]. 控制与决策，2006，21 (4)：361 - 366.

[14] 贾志成，吴小伟，茹煜. 直升机植保技术研究综述 [C]. 威海：新形势下林业机械发展论坛，2010.

[15] 张国庆. 农业航空技术研究述评与新型农业航空技术研究 [J]. 江西林业科技，2011 (1)：25 - 31.

[16] 薛新宇，兰玉彬. 美国农业航空技术现状和发展趋势分析 [J]. 农业机械学报，2013，44 (05)：194 - 201.

[17] LAN Y B, THOMSON S J, HUANG Y B. Current status and future directions of precision aerial application for site-specific crop management in the USA [J]. Computers and electronics in agriculture, 2010, 74 (1): 34 - 38.

[18] DALAMAGKIDIS K, VALARANIS K P, PIEGL L A. On integrating unmanned aircraft systems into the National Airspace System: issues, challenges, operational restrictions,

certification and recommendations [S]. Springer, 2009.

[19] Model Aeronautical Association of Australia. Manual of procedure [S/OL]. [2004-05-09]. http://www.maaa.asn.au/maaa/mop.html.

[20] HUANG Y, HOFFMANN W C, LAN Y. Development of an unmanned aerial vehicle-based spray system for highly accurate site-specific application [C]. Rhodes Island: American Society of Agricultural and Biological Engineers, 2008.

[21] HUANG Y, HOFFMANN W C, LAN Y, et al. Development of a spray system for an unmanned aerial vehicle platform [J]. Applied engineering in agriculture, 2009, 25 (6): 803-809.

[22] KILROY B. Aerial application equipment guide 2003 [M]. Washington: USDA Forest Service, 2003: 59-62, 143-147.

[23] HUANG Y, LAN Y, GE Y, et al. Spatial modeling and variability analysis for modeling and prediction of soil and crop canopy coverage using multispectral imagery from an airborne remote sensing system [J]. Transactions of the ASABE, 2010, 53 (4): 1321-1329.

[24] 石媛媛. 基于数字图像的水稻氮磷钾营养诊断与建模研究 [D]. 杭州：浙江大学，2011.

[25] 中华人民共和国农业农村部. 中国农业机械化年鉴 [M]. 北京：中国农业科学出版社，2015.

[26] 中华人民共和国财政部. 2014 年农机购置补贴政策取得实效 [EB/OL]. (2015-02-12) [2017-9-18]. http://nys.mof.gov.cn/zhengfuxinxi/bgtGongZuoDongTai_1_1_1_1_3/201502/t20150212_1191989.html.

[27] 陈自业. 国外航空在水稻栽培上的应用 [J]. 黑龙江农业科学，1980 (4): 56-57.

[28] 汪沛，胡炼，周志艳，等. 无人油动力直升机用于水稻制种辅助授粉的田间风场测量 [J]. 农业工程学报，2013，29 (3): 54-61.

作者简介

罗锡文，中国工程院院士，教授，博士生导师。曾任华南农业大学副校长。现任华南农业大学教授、南方农业机械与装备关键技术教育部重点实验室主任，农业农村部水田农业机械装备重点实验室主任，是华南农业大学农业机械化工程国家重点（培育）学科带头人，农业机械化及其自动化国家特色专业、农业机械学国家精品课程和农业机械学国家级教学团队负责人，兼任中国农业机械学会理事长、中国农业工程学会名誉理事长、科技部中国农村技术开发中心总体专家组智能农机装备专项组长、全国农业航空产业技术创新战略联盟理事长、全国农机化科技创新专家组组长、农业农村部主要农作物生产全程机械化推进行动专家指导组组长、国家水稻机械化产业体系专家等。长期从事农业工程教学、科研和工程实践，研究方向包括水稻精量直播技术与机具、农田精准平整技术与机具、农业机械导航与自动作业技术与装备、农业航空关键技术和农情信息快速获取技术与装备，在农业机械技术创新、农业工程学科建设、创新型工程科技人才培养和科技发展战略研究等方面做出了重大贡献。首创同步开沟起垄施肥水稻精量穴直播技术体系，研制成功的水稻精量穴直播机和水田激光平地机居国际领先水平，在国内首次成功研制无人驾驶水稻插秧机和直播机、无人驾驶棉花播种机和拖拉机。他积极推进中国农业航空事业的发展，发起并组织成立了“国家农业航空产业技术创新战略联盟”。获国家技术发明奖二等奖1项，省部级科技奖励15项；发表学术论文350余篇，主编专著教材等6部，授权发明专利70余件；获国家教学成果二等奖2项，广东省教学成果一等奖5项，二等奖2项。培养的研究生1人获全国优秀博士学位论文奖，1人获全国优秀博士学位论文提名奖，1人获广东省优秀博士后。

曾被评为国家级教学名师、全国教育系统劳动模范、农业农村部中青年有突出贡献专家、广东省优秀共产党员、南粤教书育人优秀教师和全国优秀农业科技工作者。

基于天空地综合遥感观测技术的国家粮食安全研究

唐华俊　吴文斌①

1　引言

“民以食为天。”粮食安全直接关系到国家安全、社会稳定和经济发展，及时、准确地掌握我国农作物种植面积和产量等关键粮食生产信息，对政府宏观决策、科学指导农业生产、防灾减灾、粮食安全预警和农产品贸易等具有重大意义。[1]传统的统计方法通过地面采集的方法获取某一行政区域的粮食生产动态信息，然后逐层汇总上报。该方法应用于大范围农情信息监测时耗费人力、物力和财力，具有时间滞后性；而且统计汇总易受人为因素的干扰。随着空间技术的不断发展，新兴的遥感技术因高时效、宽范围和低成本的优点被广泛应用于对地观测活动中。[2]不同的时间、空间、光谱、辐射分辨率，多角度和多极化的遥感卫星不断涌现，对地观测探测能力不断增强，为大范围的农情信息监测提供了新的科学技术手段。[3]然而，我国地形多样，多云多雨天气频发，种植制度复杂和农业生产具有高度动态性，使得利用遥感技术进行区域农情信息监测面临许多重大技术难题，单一传感器或单一遥感平台的对地观测在实际应用中存在较多局限性。[4]因此，在有关国家主体科研计划和农业农村部专项项目的支持下，经过10多年的技术攻关，综合天基、空基和地基观测，建立了天空地协同遥感观测系统，并应用于我国农情遥感监测与信息服务，为国家农业主管部门提供了大量决策信息。

2　天空地综合遥感观测技术

2.1　总体框架

图1是天空地综合遥感观测系统的总体框架。图中所标示的“天”是卫星遥感观测，具有区域范围大和空间连续性的特点，是区域农情遥感监测的信息主体。图中所标示的“空”是航空遥感观测，包括有人机和无人机遥感平台，具有高精度和时间连续性的特点，可以补充遥感信息的缺失，是中小尺度农情遥感监测的重要手段。图中所标示的“地”是物联网和互联网结合的地面传感网，具有实时观测和快速传输的特点，

① 唐华俊，中国农业科学院副院长；吴文斌，农业农村部农业遥感重点实验室副主任。

提供地面真实信息，服务天空平台精度验证。[5]通过天空地协同的遥感观测系统，进行农情信息的采集、融合、同化与集成应用，可以突破农情遥感监测数据时空不连续的关键难点，显著提高信息获取保障率，实现对农情信息全天时、全天候、大范围、动态和立体监测与管理，为农业生产管理和资源优化利用提供信息支撑。[6]

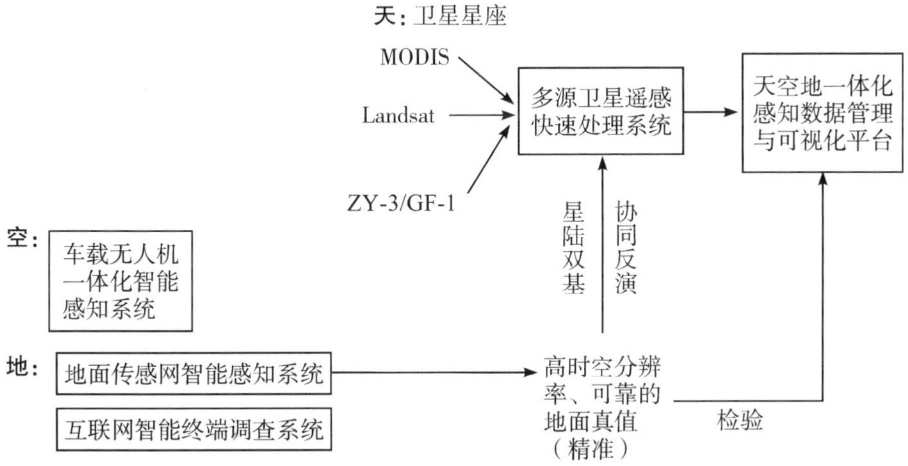

图 1　天空地综合遥感观测系统的总体框架

目前，我国建立了农业农村部唯一的卫星遥感数据接收系统，稳定获取 Landsat、HJ、MODIS、NOAA 和 FY 等国内外不同时空分辨率的卫星遥感数据。在国家高分科技专项的支撑下，近年建立的国家高分数据农业分中心实现了 GF-1 和 GF-2 国产高分卫星数据的接收、处理、分发。高分卫星数据覆盖范围大、空间分辨率高已经成为农情监测主体业务的主要信息源，提高了国产卫星遥感数据的使用率，降低了数据购买成本。[7]

我国建立了基于无人机和地面车载平台的地面样方农情信息采集系统。地面样方抽样调查是进行卫星遥感地块信息提取的重要组成部分，其调查的精度和效率直接影响着遥感地块信息提取工作。小型无人机由于其操控简便、数据获取效率高等优点，有利于快速有效地获取更大面积的地面样方数据，有效解决了传统人工调查方法在效率方面的不足。地面移动车载遥感平台搭载定位系统、相机和三维激光扫描仪等传感器可以在无人机获取数据的同时获取大量的农田影像[5]；同时，车载平台可以作为无人机运输、控制平台，也可以承担数据存储、预处理、远程传输等功能，实现空地一体化地面样方信息快速智能获取。

我国引入物联网技术，通过自主设计开发的核心硬件（传感器节点）、多功能集成的单点数据采集平台、稳定高效的网络节点布设，建立了覆盖全面、层次多样的地面获取网络，包括遍布全国的野外台站体系、地面网点采样县，为农作物遥感监测研究与应用提供了真实的数据支撑。同时，我国建立了基于移动互联网的农户种植调查系统，通过个体农户自行对经营地块进行确认，获取个体农户生产决策信息，弥补地图/遥感影像仅能反映地块自然属性特征的不足，实现“人—地”信息的有效结合。[8]

2.2 系统组成和关键技术

天空地协同遥感观测系统利用遥感网、物联网和互联网三网融合获取全面的农田和环境信息，实现农情信息快速、自动感知、采集、传输、存储和可视。其关键技术包括：多源卫星遥感影像快速处理系统、搭建车载无人机一体化智能感知系统、组建基于地面传感网的智能感知系统、基于互联网智能终端的调查系统，以及天空地一体化感知数据管理与可视化平台5大系统。

多源卫星遥感快速处理系统利用高效的金字塔算法、高精度图像配准算法、退化函数提取算法、图像恢复算法和基于深度学习的超分辨率重建算法，包含控制点管理模块、预处理模块、高级处理模块和影像操作等主要模块。该系统实现了多源高分辨率卫星遥感进行快速浏览，辐射校正，几何校正，多光谱和全色影像的融合、镶嵌、裁剪，图像恢复和超分辨率重建等处理功能，为天空地一体化农情信息管理和可视化平台提供可靠支撑。

车载无人机一体化智能感知系统利用遥感技术、地理信息系统技术、全球定位技术、互联网和大数据等先进技术，基于车载遥感平台，集成了三维地理信息与任务规划系统、无人机遥感获取系统、车载遥感获取系统、多平台融合的农情监测快速处理系统和数据远程传输系统。该系统为遥感地面样方调查、农情监测、农业灾害核查等提供了一套地面及近低空、有效的全流程移动式遥感解决方案。

地面传感网智能感知系统实现无人值守的农情信息自动、连续和高效获取。通过物联网和传感器技术直接获取农田环境参数，包括空气温湿度，风速风向，光合有效辐射强度，降雨、土壤分层温湿度，水温、水深、作物叶面积指数和实时视频等指标参数，感知作物生长过程情况，用来辅助决策田间管理措施和预测其产量等。

互联网智能终端调查系统通过手机、平板电脑、移动电脑等终端平台，基于地图、遥感影像等空间信息，进行个体农户经营地块确认，并针对地块进行生产决策信息（包括权属、作物管理、生产投入、农业产出等）采集，为农业大数据研究与应用提供基础数据支撑。[8]

天空地一体化综合观测数据管理与可视化平台通过搭建云平台，实现多源遥感数据、车载无人机数据、地面传感网数据、历史数据以及其他空间数据的统一管理、显示、存储。基于这些大数据和云计算技术，我们利用深度学习等智能识别算法来实现农情信息自动化数据处理和分析，解决当前人工处理地块数据的低效率问题。

3 天空地遥感综合观测技术的应用实践

3.1 应用领域

天空地综合遥感观测系统的农业应用领域广泛，覆盖大田作物、设施农业、农牧场和水产养殖等，尤其大田农作物遥感监测，因其对国家粮食安全的重要性而成了系统应用的首要领域。[9]基于天空地综合观测的农作物遥感监测内容总体上包括现状调查和变化监测两个方面。现状调查主要利用天空地协同遥感观测来获取某一时间或瞬时下以空间和质量属性为主的农情状态信息，包括耕地和农作物的类型、数量、分布和质量，以及农田土壤肥力、水分和污染状况。[2]变化监测主要利用天空地协同遥感来获取区域农情在特定时间段内的发生变化的位置、分布、范围、类型等动态信息，包括耕地和农作物生产全过程的势情、墒情、灾情和病情变化，农业旱涝灾害解析，农作物种植结构变化，农田生态环境退化和全球变化农业影响评估分析等。[10-14]

天空地综合遥感观测系统的应用服务包括宏观和微观两个层面。宏观应用是充分发挥协同对地观测的范围广、速度快和效率高的技术优势，在国家或区域层面进行农情监测、工程监管和信息服务，其服务对象多为政府、事业单位等资源环境监管和决策部门。例如，国家级农业遥感监测业务化运行系统从 2002 年开始运行，成为农业农村部农情会商的主要信息源之一，每年向农业农村部提交监测报告 100 余期，20 余期被国务院办公厅《昨日要情》、农业农村部《每日要情》等采用，为国家粮食安全、粮食进出口贸易等宏观决策提供了大量准确的科学数据支撑。微观应用是综合集成遥感、地理信息系统、导航定位系统、农业专家系统和智能装备系统，以地块为单元的精准或智能农业管理应用，实现基于空间变异的定位、定时、定量和精准操作，其服务多面向农业生产一线的农户、农场主和涉农企业。目前，河南省鹤壁、吉林省长春和榆树等地已经建设了智慧农业示范工程，为指导当地农业生产、推动农业信息化发挥了重要作用，产生了显著的经济、社会和生态效益。

3.2 全国主要粮食作物种植面积本底详查

农作物种植面积是作物估产的基本要素，快速准确地掌握中国主要粮食作物种植面积及其空间分布，对于辅助科学决策和政策制定、调整种植结构和确保国家粮食安全具有重要意义。从 1998 年开始，我国农业农村部陆续开展了 7 大农作物种植面积遥感监测业务化运行，但以空间抽样调查为主，得到抽样区单一作物的空间分布信息以及整个监测区域的目标作物面积统计信息，未形成全国或主产区的农作物空间分布本底图。从 2008 年开始，农业农村部开始利用天空地综合观测数据进行全覆盖，开展了主要农作物种植面积本底详查，先后利用 5 ～ 30 米空间分辨率的遥感数据完成了全国水稻、冬小麦和玉米等大宗粮食作物种植面积本底详查。2013 年以来，随着 GF-1 和 GF-2 等“高分”系列卫星的发射，特别是 GF-1 卫星宽视场 16 米空间分辨率的多光谱数据，相比 Landsat 等中分辨率卫星影像，提供了更加丰富的纹理细节和空间结构信息。同时，4 天的重访周期为大范围农作物种植面积本底详查提供了更多有效数据源。[1]

在技术层面，我国建立了空地多平台融合的地面数据采集和信息解析、多源数据协同和多特征量优化组合的多作物智能分类关键技术，实现“数据获取—作物制图—信息服务”等流程化和集成化；利用无人机和车载平台进行目标作物地面样方信息的采集，为基于卫星的区域作物识别和分类提供基础地面信息支撑。无人机地面样方采集首先根据农作物类型、面积数量及空间分布，进行地面样方布设与优化，以保证样方数量足够和空间分布均匀，满足高精度农作物遥感制图和精度评价需求；其次，利用车载高精度差分 GPS，通过无人机和车载平台采集数据的相互匹配和联合定位，实现地面信息的协同采集与高精度处理；最后，利用车载平台获取的影像数据，对无人机影像进行识别和分类，获取地面样方中各类作物的面积及分布，保障地面样方信息获取的效率和可靠性。[15]卫星观测影像是农作物种植面积本底调查的主要数据源，覆盖全国或目标作物主产区，在空地获取的地面样方信息支持下，利用作物独特的光谱反射、时间和空间特征，构建多个目标作物识别的最优遥感特征量，实现目标作物的高精度识别和构建空间分布制图。[16-18]目前，我国首次完成了全国第一张 1∶50000 的水稻、小麦、玉米等作物真实种植空间分布图，总体精度达到 97.5%，填补了我国主要农作物大比例尺空间分布数据的空白，为农作物种植面积监测和产量估测提供了科学准确的本底数据。[19]

3.3 农业生产全过程智能管理与决策

21 世纪以来，人类全面迈进了以互联网为中心的信息技术时代。随着物联网、大数据、云计算和移动互联网等新一代信息技术迅速发展，农业信息化正从传统的数字化、网络化向智能化、智慧化的高端方向发展。智慧农业将新兴的物联网、互联网和云计算等信息技术深入应用到农业生产、加工、经营、管理和服务等全产业链环节，实现精准化种植、互联网化销售、智能化决策和社会化服务。在智慧农业快速发展的新形势下，天空地综合遥感观测系统的作用更加凸显，尤其在大田作物智慧施肥、智慧灌溉、智慧植保和智慧气象等农业生产智能管理与决策中的应用日益深入。

借助于地面传感网，利用传感器实时采集农业生产现场的温湿度、光照、二氧化碳浓度等农田环境参数，利用数码相机、视频监控设备获取农作物的生长状况等信息，远程监控农作物和农田环境；将采集的参数和获取的信息进行数字化转换，经传输网络实时上传到智能管理系统中，建立农田土地管理、土壤数据、自然条件、作物苗情、病虫草害发生发展趋势和作物产量等的点位观测数据库。[20]同时，利用同步高精度卫星遥感数据反演或提取区域尺度农田和作物参数，[21,22]并进行空间信息的地理统计处理、图形转换与表达等，为分析农田差异性和实施调控提供处方信息。综合利用作物生产管理与长势预测模拟模型、投入产出分析模拟模型和智能化农业专家系统，在决策者的参与下，根据农田产量的空间差异性，[23,24]进行原因诊断，提出科学处方，生成田间作物管理处方图，精确地遥控农业作业设施（如远程控制节水浇灌、节能增氧等），实现精准化、智能化的农业生产操作与管理。

4 讨论与展望

天空地遥感综合观测系统农业应用的基本流程主要包括遥感数据的筛选、遥感图像预处理、遥感特征参量反演、时空信息提取、模型构建和校正、质量评估和数据集成与汇总等。目前，还有很多核心科学问题尚未得到系统的解决。

首先，天空地协同和立体遥感观测能力有待加强。目前，天空地协同观测对农业资源环境监测应用的满足度还不高，卫星和传感器参数设计未能充分满足农业的特有需求。关键作物生长期与关键农事管理节点需要微波遥感全天候遥感观测数据的获取；土壤定量遥感、作物品种与品质监测、病虫害遥感监测等需要高光谱遥感数据；作物生理与生长状态监测需要荧光遥感、偏振遥感等新型遥感器应用；天空地多源观测数据的融合与同化理论和技术方法需要完善。

其次，人工智能与大数据等支持下的信息智能提取和挖掘核心技术仍待突破。无论是土地利用类型、作物种类的分类识别，还是作物生长状态和环境要素的定量遥感，都是非常复杂的认知过程。由于遥感数据本身波段间的相关性、遥感器设计波段的有限性，以及地物同物异谱、异物同谱的光谱复杂性，遥感信息提取和智能挖掘具有病态问题，存在很大的不确定性。人工智能与大数据技术的发展，为农业资源环境信息反演、提取与应用提供了崭新的技术途径。

最后，天空地协同遥感观测的应用范围和应用领域需要进一步拓展。遥感观测与导航定位、互联网、物联网、大数据等技术的融合，与农学领域的其他学科交叉结合，可以从方法学上推动自身学科发展，同时跨学科应用也将拓展应用领域，如进一步推进天空地协同观测在精准农业、作物育种表型、农业保险监测与评估、农业工程监测和农业政策效果评估等方面的应用深度发展。

参考文献

[1] 陈仲新，任建强，唐华俊，等. 农业遥感研究应用进展与展望 [J]. 遥感学报，2016，20（5）：748－767.

[2] 唐华俊，吴文斌，余强毅，等. 农业土地系统研究及其关键科学问题 [J]. 中国农业科学，2015，48（5）：900－910.

[3] 唐华俊，吴文斌，杨鹏，等. 农作物空间格局遥感监测研究进展 [J]. 中国农业科学，2010，43（14）：2879－2888.

[4] 胡琼，吴文斌，宋茜，等. 农作物种植结构遥感提取研究进展 [J]. 中国农业科学，2015，48（10）：1900－1914.

[5] SHI Y，JI S P，SHAO X W，et al. Framework of SAGI agriculture remote sensing and its perspectives in supporting national food security [J]. Journal of integrative agriculture，2014，13（7）：1443－1450.

[6] 宋茜，周清波，吴文斌，等. 农作物遥感识别中的多源数据融合研究进展 [J]. 中国农业科学，2015，48（6）：1122－1135

[7] ZHOU Q B，YU Q Y，LIU J，et al. Perspective of Chinese GF-1 high-resolution satellite data in agricultural remote sensing monitoring [J]. Journal of integrative agriculture，2017，16（2）：242－251.

[8] YU Q Y，SHI Y，TANG H J，et al. eFarm：a tool for better observing agricultural land systems [J]. Sensors，2017（17）：453.

[9] WU W B，YU Q Y，VERBURG P，et al. How could agricultural land systems contribute to raise food production under global change [J]. Journal of integrative agriculture，

2014, 13（7）：1432 -1442.

[10] 黄青，唐华俊，吴文斌，等. 农作物分布格局动态变化的遥感监测：以东北三省为例［J］. 中国农业科学，2013，46（13）：2668 -2676.

[11] 黄青，唐华俊，周清波，等. 东北地区主要作物种植结构遥感提取及长势监测［J］. 农业工程学报，2010，26（9）：218 -223.

[12] SUN J, WU W B, TANG H J, et al. Spatiotemporal patterns of non-genetically modified crops in the era of expansion of genetically modified food [J]. Scientific reports, 2015, 5: 14180.

[13] LI Z G, LIU Z H, ANDERSON W, et al. Chinese rice production area adaptations to climate changes, 1949 -2010 [J]. Environmental science and technology, 2015, (49): 2032 -2037.

[14] LIU K, ZHOU Q B, WU W B, et al. Estimating the crop leaf area index using hyperspectral remote sensing [J]. Journal of integrative agriculture, 2016, 15 (2): 475 -491.

[15] 刘佳，王利民，滕飞，等. Google Earth 影像辅助的农作物面积地面样方调查［J］. 农业工程学报，2015，31（24）：149 -156.

[16] HU Q, WU W B, SONG Q, et al. Extending the pairwise separation index for multi-crop identification using time series MODIS images [J]. IEEE transactions on geoscience and remote sensing, 2016, 54 (11): 6349 -6361.

[17] HU Q, WU W B, SONG Q, et al. How do temporal and spectral features matter in crop classification? [J] Journal of integrative agriculture, 2017, 16 (2): 324 -336.

[18] 刘佳，王利民，杨福刚，等. 基于 HJ 时间序列数据的农作物种植面积估算［J］. 农业工程学报，2015，31（3）：199 -206.

[19] 唐华俊，周清波，刘佳，等. 中国农作物空间分布高分遥感制图：小麦篇［M］. 北京：科学出版社，2015.

[20] 邹金秋，周清波，杨鹏，等. 无线传感网获取的农田数据管理系统集成与实例分析［J］. 农业工程学报，2012，28（2）：142 -147.

[21] 夏天，吴文斌，周清波，等. 冬小麦叶面积指数高光谱遥感反演方法对比［J］. 农业工程学报，2013，29（3）：139 -147.

[22] 刘轲，周清波，吴文斌，等. 基于多光谱与高光谱遥感数据的冬小麦叶面积指数反演比较［J］. 农业工程学报，2016，32（3）：155 -162.

[23] 黄健熙，武思杰，刘兴权，等. 基于遥感信息与作物模型集合卡尔曼滤波同化的区域冬小麦产量预测［J］. 农业工程学报，2012，28（4）：142 -148.

[24] HUAN J X, SEDANO F, HUANG Y B, et al. Assimilating a synthetic Kalman filter leaf area index series into the WOFOST model to improve regional winter wheat yield estimation [J]. Agricultural and forest meteorology, 2016, 216: 188 -202.

作者简介

唐华俊，博士，研究员，农业土地资源遥感专家，四川省阆中市人。现任中国农业科学院院长、比利时皇家科学院（海外）通信院士、中国工程院院士。长期从事基于遥感技术的农业土地资源合理利用、农作物种植面积空间分布和结构变化研究。在传统耕地资源研究基础上，开拓耕地内部的农作物空间格局研究。发展了农作物遥感监测系统，科学监测农作物播种面积、种植区域及产量；创建了系列空间模型，定量解析了过去我国主要农作物种植面积空间分布和结构变化过程及规律；建立了耦合自然和社会经济因子的综合模型，模拟未来农作物空间分布变化趋势及其对我国粮食安全的影响。先后获得国家科技进步二等奖 2 项，发表论文 220 余篇，出版著作 10 部。

吴文斌，博士，现任中国农业科学院农业资源与农业区划研究所研究员，兼任中国农学会农业信息分会副主任委员兼秘书长、中国卫星导航定位协会农业农村专业委员会副主任、中国自然资源学会青年工作委员会副主任、《中国农业信息》副主编等。主要从事农业土地系统领域的科学研究，综合利用遥感、地理信息系统、空间模型、地统计学等多学科技术方法，开展农业土地系统智能感知、时空变化监测与综合效应评估等方面研究，尤其在大田种植和果园生产两个领域取得了重要研究进展。相关研究成果被全球土地计划（Global Land Programme）纳入重要案例全球共享，被新华社、《光明日报》《科技日报》《中国科学报》等媒体报道。

对从中国精准农业到智慧农业的几点思考

刘经南　高柯夫①

农业是国民经济的基石，其进步程度反映了国家文明的发展阶段。中国以仅占世界9%的可耕地面积和6%的淡水资源养育了世界22%的人口[1]，其中农业科技的创新和现代化发挥了关键作用，主要农作物耕种收综合机械化水平达到63.8%，农业科技进步贡献率由新中国成立初期的20%提高到目前的56%以上[2]。

当前，世界正处在以智慧为特征的新经济的萌发时期，传统农业也从信息农业时代迈进精准农业时代，互联网、大数据、云计算、3S（GNSS/RS/GIS）、物联网及人工智能等技术与农业生产、经营、管理和服务等业务的深度融合，也正引领农业产业格局向智慧农业过渡。

信息农业是基于互联网发展起来的，以实现农业信息资源的高度共享；精准农业是基于移动互联网和GNSS网络实时动态（real-time kinematic，RTK）位置服务网以及卫星遥感，特别是低空和无人机遥感并与施肥撒药等技术发展起来的，按照时空特异，定点、定时、定量地实施一整套现代化农事操作技术与管理[3]。因此，作为智慧经济形态在农业中的具体表现，智慧农业与信息农业以及精准农业在内涵外延、基础设施、技术环境和耕作理念等方面的区别是一个值得思考的问题。

1　从精准农业到智慧农业的需求变化

精准农业发源于美国，始于GPS技术在农业中的应用。此后，遥感监测系统（remote sensing system，RSS）、农田信息采集与环境监测系统、地理信息系统（geographic information system，GIS）、作物生产管理专家决策支持系统、变量投入技术（variable rate technology，VRT）、物联网和无人机等技术与农机融合，根据农地每一操作单元的具体条件，精细准确地调整优化农业投入和相应管理措施，以获取高产和最优经济效益，同时保护农业生态环境、保护土地等农业自然资源。

精准农业重点在于微观层面的“精准”，集中体现在大田耕种作业中，通过实时动态地确定作业对象和作业机械的空间位置，并将此信息转变为地理信息系统能够贮存、管理和分析的数据格式，凭借3S、物联网及人工智能等技术手段，对耕种过程进行精

① 刘经南、高柯夫：武汉大学卫星导航定位技术研究中心。

准控制，对作物长势、受灾等各方面的情况进行精准监测，再根据监测情况，精准调节耕作投入，实现耕作、灌溉、施肥施药、播种及收获等农业生产全过程的精细化管控。

按照农作物栽培对水平方向和垂直方向定位精度要求的不同，对精准化的大田管理、精准播种和农田平整等环节都有不同的精度要求。例如，依据美国的精密农业标准，大田管理的水平定位精度为 10 ～ 12 厘米，垂直定位精度为 15 ～ 28 厘米；精准播种的水平定位精度为 1. 2 ～ 2. 5 厘米，垂直定位精度为 2 ～ 4 厘米；农田平整的水平定位精度为 10 ～ 12 厘米，垂直定位精度为 1. 2 ～ 2. 5 厘米。

美国是世界上最早实施精准农业的国家。1993 年，美国将 GPS 系统技术应用到农业生产领域，在明尼苏达州农场进行了精确农业技术试验，用 GPS 指导施肥的作物产量比传统平衡施肥作物产量提高 30% 左右[4]，使美国以用仅占人口总数 2% 的农民养活了 3 亿多的美国人，并且成为全球最大的农产品出口国[5]。

目前，中美精准农业在精细化管理程度上还存在不小差距。例如，美国的精准农业大田管理最小面积是 1 平方英尺（0. 093 平方米），而中国的管理单位则以亩计（666. 67 平方米），两者差距约 7168 倍。中国有 20 亿亩耕地，按照美国的标准管理，将会有约 14. 3T 个管理单位，是典型的农业大数据。如果这些管理单位全部得到管控，未来我国精准农业的市场增长潜力将会十分巨大。

智慧农业的重点在于宏观层面的“智慧”。智慧农业作为智慧经济形态在农业中的具体表现，不局限于生产环节，而是深度融合云计算、大数据、人工智能和物联网等新技术，贯穿在农业产前、产中、产后全产业链条，覆盖农业生产、管理、营销和服务等各个环节的高级农业形态。与精准农业相比，智慧农业涵盖的范围更广，不仅包括设施农业、安全追溯、种植业作物及养殖业动物的病虫害预报与预警、农业电子商务等涉农领域，以及通过“互联网 + 现代农业”的手段，实现全程信息服务与指导，还包括创新商业运营模式，如共享农业，外延可扩展到农业经济范畴。因此，智慧农业在耕种作业精准化、基础设施智能化和产业发展现代化等领域，存在更多、更高的应用需求。

1. 1　从精确辅助驾驶到智能无人驾驶

相关全球研究报告显示，精准农业市场正以 13. 36% 的年复合增长率快速发展。2013 年，全球精准农业市值约 19. 9 亿美元，预计 2020 年将达到 45. 4 亿美元[6]，市场前景广阔。2015 年 8 月，普渡大学发布的美国精准农业相关技术应用情况调研显示[7]，GPS 导航、自动驾驶、基于 GPS 的喷药控制、GPS 物流和无人机等在精准农业相关技术的应用较为广泛（见图 1），表明以 GNSS 为主导的精确辅助驾驶技术正逐步成熟。

目前，农机车辆导航对定位精度的需求按照不同的生产环节可分为：普通精度（适合大田作业）12 ～40 厘米，较好精度（适合施肥施药）4 ～ 12 厘米，以及高精度（适合播种、中耕、收获）2 ～ 4 厘米。随着农业人口的减少，农田规模的扩大，农时的全气象条件、全天候延展，高精度定位导航技术的普及，对无人农机具的自动驾驶需求将会日益迫切。通过预先输入的农田形状和面积数据，借助高精度 GNSS 确定农机具的位置并规划路径，由农机具自动控制方向盘和耕作装置等，实现 24 小时无人驾驶状态下，田间耕作、施肥和播散农药等工作的自动操作与远程智能控制。国内已经有基于北斗卫星导航系统的农机自动驾驶系统面市，集高精度卫星导航、自动控制和无人驾驶于一体[8]，作业重复误差在 2.5 厘米以内，大幅减少了农机作业的重复面积。

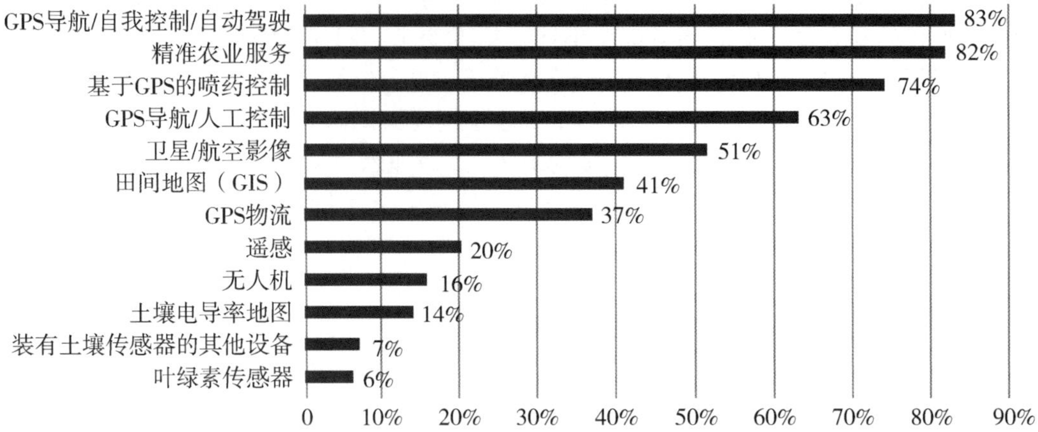

图 1　精准农业相关技术 2015 年应用情况

1.2　从精准感知到实时智能感控

随着大田农业发展达到精细化的水准，以及传感器技术的迅猛发展，农田传感器的感知对象不断扩展。传感网络可以实现自检测、自修复、自组织和自适应，从而为感知向实时化、管控向智能化的发展奠定了技术基础。

目前，世界上应用广泛的农用传感器包含温度传感器、湿度传感器、pH 传感器、溶解氧传感器、光电传感器和风速传感器等，应用领域涵盖种植环境感知、土壤温湿度检测、农药、食品安全和水资源管理等。按照感知对象分类，可分为生命信息感知技术和环境信息感知技术。其中，与生命信息感知相关的植物病虫害、草害检测与预警预报技术，作物生态信息及三维形态虚拟、模拟技术，以及作物、牲畜生产管理决策支持技术，都存在运用大数据、云计算及人工智能方法实现决策智能化的应用需求；与环境信息感知相关的土壤有害物信息快速检测技术，土壤养分、理化信息快速动态感知技术则存在感知实时化的应用需求。

我国传感器的种类目前仅为世界传感器种类的 1/10 左右。随着智慧农业的发展，我国农业传感器在农业装备、设施农业、水产养殖、品质控制和存储保鲜等领域的应用会更加广泛。例如，将传感器设备与自动驾驶农机具结合，可以采集农机具的位置、图像、视频、深松作业和侧翻报警等信息，并通过高速无线通信网络传输给系统管理服务平台，供平台对采集到的数据进行分析和挖掘，从而实现智能调度、远程测亩、实时感控、深松检测和异常报警等功能。

1.3　从信息收集分享到资源和生产资料分享

随着互联网、大数据及云计算技术与农业科技的深度融合，结合地面观测、传感器、遥感和地理信息技术等多种田间信息获取技术，精准农业涉及的各种类型的海量数据快速形成。通过农业生产环境、生产设施和动植物本体感知数据的采集、汇聚和关联分析，以信息分享为主的农业大数据对提高精准农业的生产管理、指挥调度能力，农机作业质量的远程监控能力，作物种植面积、生产进度、农产品产量的关联监测能力起到了有力的支撑作用。[9]

精准农业采集分享的农业大数据，作为智慧农业的神经系统，将不局限于作为生产要素参与农业生产，还将与农机、专家、领域知识等资源和生产资料深度融合，参与生产、经营、服务、贸易等农业产业链的相关活动中，推动新的农业发展模式、农业生态和农业业态的形成。因此，大数据的共享是非常重要的。农业领域设备一般为产业链上需要用到的大型机器，例如，旋耕机、播种机、收割机和烘干机等都是农忙时才会用到，其他时间只能闲置。以有形生产资料分享的农业共享经济模式，不但可以提高设备利用率，还将大幅降低生产成本。此外，我国农村土地流转政策一旦落地，由此带来的土地生产资料的共享，将会真正激活规模化农业生产模式，充分利用有限的土地资源，释放更多的农业人口；同样，人力资源的共享，根据农业农忙和农闲的季节规律，可以实现忙时务农，闲时进城务工的格局，从而缓解人口老龄化带来的农业人口不足、人力成本过大的问题。以农技共享为代表的无形生产资料分享的农业共享经济模式，通过线上学习和线下实践的方式，将加快农业科技的普及，让更多的农民分享到智慧农业的信息化发展成果。另外，也只有通过农情大数据的共享，才能找到同类农情数据不同时期、不同地域之间的关联，才能发掘不同类农情数据之间的关联和因果关系，从而通过基于云计算的深度学习，实现更高阶段的农业智慧。

2　农业智能感知与监控网及其应用

2016 年 1 月 27 日，《中共中央国务院关于落实发展新理念加快农业现代化实现全面小康目标的若干意见》[10]提出：“大力推进‘互联网 +’现代农业，应用物联网、云计算、大数据、移动互联等现代信息技术，推动农业全产业链改造升级。”这表明农业发展进入了新的历史阶段。

这个新的历史阶段就是智慧农业的阶段。这个阶段的第一步就是要建立能实时或定时感知和监控不同地块、地域和地区的农情变化的智能感知网络。农业智能感知与监控网的提出，正是基于“互联网 +”和物联网技术，通过各类传感平台和传感器技术，实时智能感知农业生产现场的光照、温度、湿度、土壤肥度、病害、污染指数和含水量等参量以及农产品生长状况等参数，通过信息化形成大数据，借助计算协同和精准控制，实现远程监管、预警、决策及智能控制耕作和智慧管理等生产经营活动，完成智慧农业的数据化、网络化和智能化，以推动农业产业结构升级、产业组织优化和产业创新方式变革。

2.1　农业智能感知与监控网的基础设施架构

发达国家现代农业的技术研究取得很大成就，为世界农业发展提供了重要科技支撑。根据发达国家的技术研究进展，目前农业传感器技术、农业机器人、农业遥感和农业大数据等前沿性技术已成为发达国家抢占农业农村信息技术制高点的重要领域。农业专用传感器和仪器仪表技术发展迅速，适应农业环境、高精度、低成本、易维护的农用传感器得到攻克；以 GNSS 地基增强系统为代表的高精度导航定位技术为农机具、无人机的自动驾驶提供了时空位置基准；结合多种遥感手段，天空地一体化遥感技术手段智能获取农业大数据，全面监测作物的长势，指导作物生产管理精细化成为精准农业的研究重点；RFID 技术已广泛应用于农产品质量安全溯源；知识模型技术广泛应用于农业生产决策；多种电子监视、控制装置已应用于复杂的农业装备，土地精细平整设备、变量作业装备、农业机械自动导航和联合收割机等高度智能化农机装备得到熟化应用。

因此，农业智能感知与监控网的基础设施需要陆海空天网一体化架构，只有这样，才能为上述前沿技术的实施、融合和创新提供平台支撑。按照时空粒度的不同，初步设想在农业生产现场以物联网技术作为地基感知网的基础，实时感知农业作物、生产环境、农机具等设备状态，形成农业大数据源；在农业活动区域以卫星连续参考站（continuously operating reference stations，CORS）作为地基增强系统，建立农业生产活动的时空位置基准；在局域性低空和中空建立移动遥感平台，用于实时监测局域农情状态、土地确权、保险定损等；在区域性高空根据需求建立定点定时的遥感平台，实现大区作业的区域农情状态监测、病虫害监测和农作物估产等；在广域的高空区域，部署农业低轨遥感卫星以获取广域范围不同类型和分辨率的对地观测和地球遥感数据，再结合 GNSS 天基增强系统，对地球表层的运动目标提供良好的动态高精度定位导航服务，整个体系构成了具有高精度时空位置服务的农业智能感知与监控网络（见图 2）。

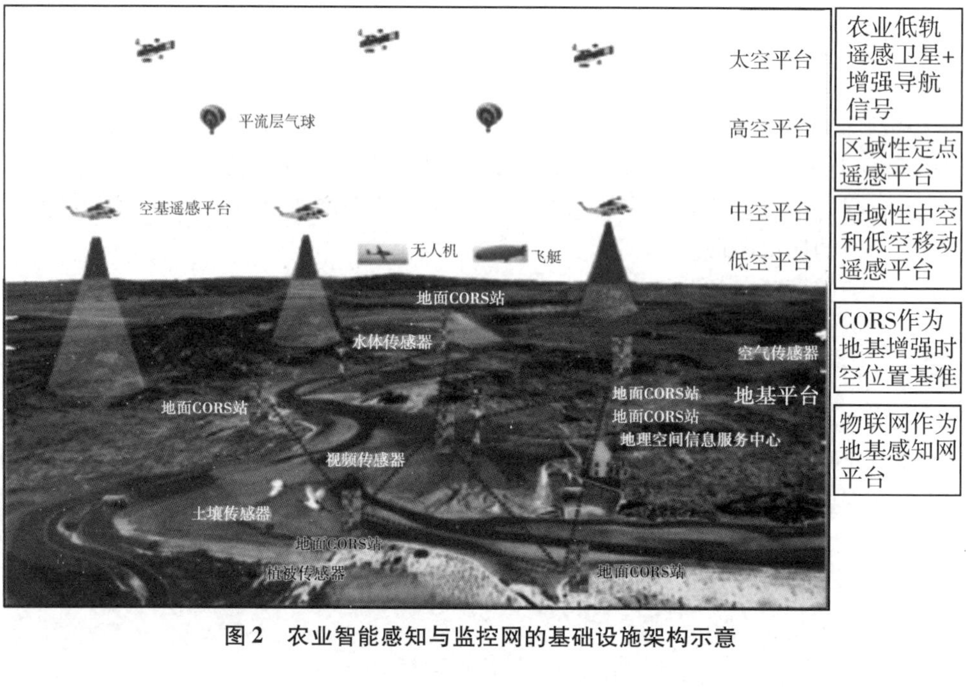

图 2　农业智能感知与监控网的基础设施架构示意

2.2　农业智能感知与监控网的关键技术

智慧农业的发展趋势使技术研发重点由信息感知向智能决策与自动化控制方向转变，不同类型技术的交叉与融合集成越来越广泛。农业智能感知与监控网的关键技术主要包括以下方面。

2.2.1　农业物联网技术

物联网作为感知基础，通过农业信息感知设备，按照约定协议，把农业系统中的动植物生命体、环境要素、生产工具等物理部件和各种虚拟“物件”与互联网连接起来，进行信息交换和通信，以实现对农业对象和过程智能化识别、定位、跟踪、监控和管理。发展符合农业多种不同应用目标的高可靠、低成本、适应恶劣环境的农业物联网专用传感器，检测环境中的温度、相对湿度、pH 值、光照强度、土壤养分和二氧化碳浓度等物理量参数，实现网络自组织和感知节点自动优化等，为实时监测和远程智能控制提供依据，帮助农民以更加精细和动态的方式认知、管理和控制农业中各要素、各过程和各系统。

2.2.2　农业大数据技术

农业大数据是大数据技术向农业领域延伸应用的一部分，包括农业自然资源与环境数据，农业生产数据（种植业生产数据和养殖业生产数据），农业市场数据（市场供求信息、价格行情、生产资料市场信息、价格及利润、流通市场和国际市场信息等），农业管理数据（国民经济基本信息、国内生产信息、贸易信息、国际农产品动态信息和突发事件信息等）。[11]农业每年产生的数据量约为 8000PB（10^{15}字节）。[12]数据是智慧农业建设的核心要素，通过大数据技术与云计算相融合，建立统一的大数据中心云平台，将为农业智能感知与监控网的用户提供实时定制的计算服务资源。农业大数据的应用是智慧决策管理和实时精准控制的基础，基于农业大数据的机器学习和深度学习是智慧决策的基础。

2.2.3　信息物理系统

《农业农村部关于推进农业农村大数据发展的实施意见》明确指出，加强对农情、植保、耕肥、农药、饲料、疫苗和农机作业等的相关数据的实时监测与分析，提高农机作业质量的远程监控能力，开展耕地、草原、林地、水利设施和水资源等数据在线采集，构建国家农林资源环境大数据实时监测网络。该检测网络的本质就是信息物理系统（cyber-physical system，CPS）。

智慧农业作为农业发展的高级阶段，趋同于工业生产方式，对远程控制以及通信计算的精准协同提出了更高的要求。CPS 通过物理设备联网，让广域或全球性分布的物理设备具有计算、通信、精确控制、远程协调和自治五大功能，以实现人类对物理世界的感知和控制在时间、空间尺度方面的延拓。农业智能感知与监控网作为典型的 CPS 网络，是分布式的异构系统，不仅包含许多功能不同的子系统，且子系统之间结构和功能各异，分布在不同的地理范围内的各子系统之间，需通过有线或无线的通信方式相互协同工作，必须掌握精准的时空位置信息，以精确推断物理位置、控制通信和协调分布式计算流程的相关活动，使计算和物理实体单元各项活动能够更紧密地互相协同配合，从而在此基础上实现广域、远程、异构的计算与控制的协同。

2.2.4 3S 技术

3S 技术是遥感技术（remote sensing，RS）、地理信息系统（geographic information systems，GIS）和全球卫星定位系统（global navigation satellite systems，GNSS）的统称，是空间技术、传感器技术、卫星定位与导航技术、计算机技术和通信技术相结合，多学科高度集成的现代信息技术。

GNSS 是利用飞行的卫星不断向地面广播发送某种频率，并加载某些特殊定位信息的无线电信号，来实现定位、导航和测时的导航定位系统。精准农业对卫星导航定位技术的应用需求，主要包括两个方面，一是为农用机具的运行控制提供高精度实时位置信息和为农田状态信息的获取提供位置信息，二是为精准农业实施变量作业提供必要条件。不同的农机作业，需要不同的定位精度，自动驾驶、变量播种、土地整理等需要厘米级定位精度，信息采集、变量控制、产量监控、作业统计等需要分米级定位精度，农机监管、调度导航、作业监测等仅需要米级定位精度。

RS 从近地空间（地面、低空、高空、太空）对地球发射某种类型的无线电波（可见光波、微波、红外波、激光等），然后接收来自地球表层各类地物的电磁波反射信息，并通过对这些信息进行扫描、摄影、传输和处理，从而对地表各类地物和现象进行远距离控测和识别。RS 可用于植被资源调查、作物产量估测、病虫害预测和自然灾害监测等方面。

GIS 是精确农业整个系统的承载动作平台和基础，各种农业资源数据的流入、流出以及对信息的决策、管理都要经过 GIS 来执行。作为精确农业的核心组件，GIS 将 RS、GNSS、专家系统、决策支持系统等组合起来，起到“容器”的作用。GIS 主要用于各种农田土地数据，如土壤、自然条件、作物苗情、产量等的管理与查询，也能采集、编辑、统计、分析不同类型的空间数据。

在农业生产中，RS 是 GIS 的一个重要数据源和强有力的数据更新手段。GIS 作为一种空间数据管理、分析的有效技术，可以为 RS 提供各种有用的辅助信息和分析手段。而 GNSS 则为 RS 和 GIS 综合系统中处理的时空数据提供实时准确的时空位置服务提供了获取手段，并作为一个重要数据源为 GIS 提供时空位置数据。智慧农业生产中，三者已发展为不可分割的整体，相互渗透，相互补充。

2.3 农业智能感知与监控网的应用

智慧农业横跨农业科技、信息技术、卫星技术、物联网、大数据、云计算和人工智能等多个领域，农业智能感知与监控网的应用将促进各领域进行深度融合，各产业相互渗透、相互促进，扩大智慧农业的产业规模。

农业智能感知与监控网用于农业自然灾害监测预报。通过建设监测站点，结合运用有关部门卫星遥感数据、无人飞机航拍数据以及 GNSS 实测数据，采集各个地区的地理位置、土壤状况、农作物种植情况和水利基础设施分布情况，以及气象灾害和病虫害发生情况等，利用不同时期自然灾害的专题数据，形成矢量数据库，输入 GIS 空间和属性数据库，做到作物病虫害的覆盖范围感知和严重程度测定。利用装载于无人机、飞机或卫星上的激光扫描雷达对城市、森林、管线等进行三维扫描，可获得 3D 立体模型。对森林进行三维扫描，还可获知其树木种类、木材储量、生长状态等参数，为森林灾害损失评估提供依据。

以农业智能感知与监控网为主要的数据采集手段，结合无人机航拍、光学与微波主被动协同的地理信息技术，建立农业风险遥感数据采集、分析和展示平台，打造“按图作业、按地管理、服务到户、防灾减损”的农业保险量化管理新模式，可以为农业保险灾害预测预报、作物面积估测和受灾评价、定点查勘、损失程度鉴定和作物产量估计等提供量化数据和精细评估。针对农业保险业务的各个主要环节，基于农业智能感知与监控网，深入地进行应用及开发，并有效整合农业防灾减损、综合服务的资源，可为农业防灾减损提供有力的保险决策支持。

对农业智能感知与监控网各环节所涉及的时空位置数据，如农田质量数据、气候预报数据、作物品种数据、耕种情况数据、畜牧位置数据、土壤条件数据和病虫害数据等方面，精密农业提出了位置精准的要求。通过遥感和精密定位，分析后建立模型，可对粮食等的产量和质量做出较为准确的估算。利用采集到的农业气象大数据，还可对区域气象历史数据建立天气模式，并将这些模式与当前的气候条件进行比较，再运用大数据分析进行天气预测。

3 对构建天空地/通导遥一体化农情智能感知网的思考

低轨农情遥感卫星导航增强服务以低轨农情遥感卫星为载体，提供区域或全球的时空基准，协同其他辅助定位导航技术以提升其抗干扰能力和可用性，利用互联网和移动互联网等通信网络，向用户提供位置、方向、速度、时间信息和时间同步服务。具体实施可由农业农村部牵头发射6～18 颗低轨农情遥感卫星，组网实现全球覆盖，每颗卫星除了主要装有可见光、短波红外遥感器以及热红外扫描仪，以获取地表温度等信息，用于对农作物长势、病虫害、水灾旱灾监测，还搭载 GNSS 信号发生器作为导航定位信号源，通过 RTK 技术，提供全覆盖的米级到厘米级实时导航增强服务，与地面共同组成天空地一体化农业智能感知和监控网。若进一步结合通信卫星功能，提高遥感卫星数据的星上快速或实时处理能力，这一地球空间信息卫星感知网络还具有快速或近实时地球遥感信息分发能力。特别是结合通信卫星将要具备的与手机直接通信的功能，就能实现遥感图像手机收发。低轨卫星星座同时还有极强感知 3D 电离层和 3D 对流层大气动态变化的能力，这一基础设施对于农情监测和认知地球全球快速和长期变化，将是革命性的变革。

以此为基础，构建 GNSS 地基增强系统，能够实现农庄（或省市域）厘米级、全国分米级、全球米级高精度实时定位导航，满足大豆等作物厘米级精度播种、分米级精度大田精细管理、米级精度拖拉机数千公里①跨区作业的调度管理的现代精密农业需求。此外，还可根据拖拉机实时移动获取的 GNSS 数据，结合地基增强基准站网分析大田土壤湿度和含水量变化。区域 CORS 网通过 GNSS 遥感大气水汽变化，不仅可用于大范围长期水汽变化监测，也可用于局域水汽变化监测，服务于天气长期变化和短期精准预报。

4 结论

我国是农业大国，而非农业强国。智慧农业是信息农业、精准农业发展的新阶段，是基于精准时空位置信息和信息物理系统对农情资源及其变化智能感知、分享、管理、决策和智能控制与调节的网络设施的农业智慧经济。对我国而言，智慧农业也是消除贫困、发挥后发优势、实现换道超车的主要途径。

智慧农业同时涉及和满足人类衣、食、住、行、智慧生存的经济形态，是唯一能实现第一产业、第二产业和第三产业一体化智慧运行的经济形态。敏锐把握从精准农业到智慧农业的需求变化，通过农业智能感知与监控网的国家和分省区市县结合的多级规划、实施和建设的协同机制，把智慧农业的应用延伸到农业生态的每一个神经末梢，将智慧农业发展为集“创意、创新、创造、创业”于资源调集、要素配置、生产、物流、销售和分享全过程链、全产业链的产业形态。

智慧是大自然进化的最高形式。因此，智慧农业必须是大自然的，自然是资源循环的、自洁的、不会产生垃圾和任何污染的，智慧农业因而也必须是资源节约的、环境友好的，也就是绿色的。智慧农业还有许多的未知，需要我们去探索、去解决。

参考文献

[1] 史舟，梁宗正，杨媛媛，等. 农业遥感研究现状与展望 [J]. 农业机械学报，2015，46 (2)：247－260.

[2] 聂英. 中国粮食安全的耕地贡献分析 [J]. 经济学家，2015，1 (1)：83－93.

[3] 曹幸穗，柏芸，张苏. 大众农学史 [M]. 山东科学技术出版社，2015.

[4] 陈桂芬，于合龙，曹丽英. 数据挖掘与精准农业智能决策系统 [M]. 科学出版社，2011.

[5] 吴海鹏. 试析美国全球粮食战略的发展历程 [J]. 世界经济情况，2013 (4)：20－25.

[6] 赵春江. 智慧农业发展现状及战略目标研究 [J]. 智慧农业，2019，1 (1)：1.

[7] 王媛，严琳. 精准农业各项技术在美国应用现状 [J]. 农业机械，2016 (4)：66－68.

[8] 吴林，施闯，姜斌. 基于北斗/GNSS 星基 PPP 增强技术的农机自动导航驾驶系统 [J]. 农机科技推广，2019 (10).

[9] 许世卫. 农业监测预警中的科学与技术问题 [J]. 科技导报，2018，36 (11)：

① 1 公里 =1 千米。

32 -44.

[10] 国务院. 关于落实发展新理念加快农业现代化实现全面小康目标的若干意见 [EB]. [2016 -01 -27]. http://www.gov.cn/zhengce/2016r01/27/content_5036698.htm.

[11] 戴旭宏, 倪玖斌. 大数据驱动乡村振兴共享共治机制研究 [J]. 大数据, 6 (2).

[12] 刘智慧, 张泉灵. 大数据技术研究综述 [J]. 浙江大学学报 (工学版), 2015, 48 (6): 957 -972.

作者简介

刘经南，大地测量与卫星导航专家，北斗/GNSS 技术应用和工程领域专家。1967 年本科毕业于武汉测绘学院五年制天文大地测量专业，1982 年研究生毕业于武汉测绘学院天文大地测量专业，获工学硕士学位。2003—2008 年任武汉大学校长，2012 年起任昆山杜克大学校长。现任武汉大学教授，博士生导师，国家全球卫星定位系统工程技术研究中心主任，中国测绘学会常务理事，国际 *GPS World* 杂志编委，国际 GPS 地球动力学服务组织协调成员，中国兵器工业集团公司双跨院士，中国北斗卫星导航系统专家委员会委员。

20 世纪 90 年代初，负责完成了中国南海诸岛 GPS 大地控制网的建立和数据处理以及国家 GPS A、B 级网的设计方案和数据处理；随后负责中国广域差分 GPS 建设方案等多项科研项目、湖北清江隔河岩大坝 GPS 形变监测系统的总体设计方案和软件开发等；在大地测量坐标系理论、卫星定位应用、软件开发和重大工程应用方面做出了一系列开创性工作；主持研发的深圳市连续运行卫星定位服务系统，推动了省级及城市连续运行卫星定位服务系统在中国的建设。先后 5 次获得国家科技进步奖、多次省部级科技进步奖和 1 次国家教委教学成果一等奖，还曾获得中国科协“全国先进科技工作者”等多项荣誉称号。

高柯夫，1981 年出生，博士，副研究员，国家卫星定位系统工程技术中心主任秘书。主要从事位置服务、智能交通及北斗应用方面的研究，发表相关 SCI/EI 期刊论文 10 余篇，授权发明专利 5 项。主持武汉市基础研发计划 1 项，参与十三五重点研发计划课题 1 项、国家自然科学基金项目 1 项、国家/行业标准制订项目 2 项、国家发展和改革委员会北斗创新发展重大工程项目 1 项、海军装备科研研究 1 项。

Agricultural Aviation and Precision Agriculture: Advancing to the Next Decade

Yanbo Huang①

1 Introduction

Precision agriculture has been a milestone to revolutionize agricultural operations since late 1980s. Precision agriculture has been established on agricultural mechanization with the integration of global positioning system (GPS), geographic information system (GIS) and remote sensing (RS) technologies. Over the thirty years precision agriculture has been developed from strategic monitoring using satellite imagery for regional decision making to tactical monitoring and control using low-altitude remotely sensed data for field-scale site-specific treatment. Agricultural aviation is a key technology that ensures the success of precision agriculture.[1] With the advances of information and electronic technologies, especially the Internet, data science, and new transducers/sensors, aerial operations to monitor and treat crop fields can be integrated in the real- or near real-time mode.[2]

According to the definition by the National Agricultural Aviation Association (http://www. agaviation. org), agricultural aviation is an industry that "consists of small businesses and pilots that use aircraft to aid farmers in producing a safe, affordable and abundant supply of food, fiber and biofuel" while "aerial application is a critical component of high-yield agriculture." Now in the United States of America there are approximately 3,600 agricultural aircrafts (87% of which are fixed-wing aircrafts and the rest are helicopters) to complete about 25% of the crop protection work (http://international-pest-control. com/3730-2/). For UAV (unmanned aerial vehicle) aerial application for crop protection, after Japan developed the Yamaha technology in 1980s,[3] China began to rapidly develop UAV plant protection technology[4]. However, Ken Giles[5] made comments that "in the U. S. right now there is no commercial use of this technology—it's strictly a research and development effort," which is still the situation up to today as he described.

Variable-rate application, a key technology for crop protection in precision agriculture,

① Yanbo Huang, United States Department of Agriculture (USDA), Agricultural Research Service (ARS), Crop Production Systems Research Unit (CPSRU), Stoneville, Mississippi, USA.

has been commercialized and ready to use in the ground-based platform, typically tractor mounted systems.[6] However, aerial variable-rate application is still on research and development effort with the challenge of sprayer control timely response to remote sensing prescription.[7-9]

This paper overviews our research projects and activities in the past ten years in aerial application technology and remote sensing, and outlooks the perspectives of agricultural aviation and precision agriculture in the next ten years.

2 USDA-ARS CPSRU Precision Agricultural Research

USDA-ARS CPSRU is located at Stoneville, Mississippi, USA in the center of Mississippi Delta, an important agricultural research and production region. CPSRU has more than ten scientists working on scientific research on weed science, crop genetics and precision agriculture. In the precision agricultural group, scientists have been focusing on research and development of aerial application and remote sensing technologies to improve the effectiveness of crop protection and production in the aspects as follows.

2.1 Aerial Application Technology

The research of aerial application technology has been conducted by scientists of USDA-ARS at Stoneville, Mississippi for almost twenty years. In the past two decades, scientists evaluated various nozzles for drift control and tested a hydraulic pump with an automatic flow controller for aerial variable-rate spray studies. Figure 1 shows the Air Tractor 402B (Air Tractor Inc., Olney, Texas, USA) with spray booms and flow control system for different nozzles.

Figure 1 Air Tractor 402B with spray booms and flow control system for different nozzles used for aerial application research in USDA-ARS CPSRU

2.1.1 Nozzle Characterization

A low drift CP flat-fan nozzle was investigated for characterization of in-swath spray deposition.[10] In the study, the CP flat-fan nozzles with selectable tips and swivel angles were evaluated with different application volumes for droplet spectra and coverage using water sensitive papers placed in the spray swath. Further a study was conducted to investigate the CP flat-fan nozzle to characterize the drift at different application altitudes with a downward nozzle angle of 30 degrees optimized from the previous study.[11] With the optimized parameters from these two studies the nozzles were used to assess the crop injury from the downwind drift of the aerially applied glyphosate.[12-14]

2.1.2 Drift Management and Control

Aerial spray drift can be caused by atmospheric stability and downwind vertical to the spray line. To avoid atmospheric-induced drift the meteorological calculation is needed to determine the atmospheric stability for temperature inversion to recommend timing for aerial applicators to conduct aerial spray. Based on the wind and temperature data acquired from different heights at a 30m tower over the crop growth season in a year the daily likelihood of temperature inversion was calculated[15] and the effect of data sampling intervals was evaluated.[16] To help aerial applicators' and farmers' field operation a web site is being developed to online recommend timing of aerial application to avoid spray drift caused by temperature inversion. Similarly an App is also being developed on smart phones and tablets for applicators and farmers to have timely recommendation.

Downwind drift was characterized by using low-drift nozzles[11] and assessing crop glyphosate injury.[13, 14] The Agricultural Dispersion Model (AGDISP) was conducted to evaluate the factors that have impacts on downwind drift.[17] AGDISP is a Lagrangian based aerial spray dispersion simulation model that models spray material movement, accounting for effects of aircraft wake effects and turbulence from both aircraft and ambient sources. AGDISP is, however, a sequential cause and effect model that could benefit practically from analysis of multiple factors to obtain a set of optimal results. An approach to modeling called the Design of Experiments (DOE) technique was first introduced by Tauguchi[18] and can be used to systematically study the influence of many factors and their interactions on an outcome. We developed a new approach to identify the main factors and interactions that have significant influence on drift of aerially applied spray using the DOE technique with AGDISP.[19] With DOE, input values of the AGDISP simulation were generated randomly with a probability distribution within predetermined ranges of the input variables. In this way, outputs of AGDISP resulted from all possible values within ranges of those input variables. Then, with the simulation data and through factorial statistics, DOE identified the impact of factors on the outcomes, such as total downwind drift and interactions among the factors. Based on the results of DOE-based AGDISP simulation, Huang et al. [20] further optimized selection of controllable variables to minimize downwind drift from aerially applied sprays using the AGDISP model through DOE. With the DOE method, several near-optimal solutions for reduction of spray drift can be determined, and one could be chosen within the constraints of the aerial applicator's spray setup and weather conditions. Field validation and appropriate sensitivity analyses of this DOE-based AGDISP simulation are needed as first steps towards promoting this method to aerial applicators.

2.1.3 Aerial Application of Spray Bio-control Agents

As mentioned above, we have evaluated CP-11 flat-fan and we also evaluated Davidon tri-set nozzles. When applying liquid tank mixes from aerial platforms, there are numerous nozzle types available with differing spray characteristics. More information is needed, however, on the ability of aerial delivery systems to effectively apply biological agents. The release of non-toxigenic A. flavus into corn fields has shown promise as a biological control agent for aflatoxin producing strains of the fungus. However, the application of a coarse granule to mature, 2 meter-tall corn is a challenge. Thus, there would be substantial advantages to a liquid formulation with necessary identification of appropriate adjuvants to disperse the highly hydrophobic spores of A. flavus. The study is undertaken for effective use of these nozzles and other nozzles in application of biological control agents, especially Afla-Guard®, a commercially available product containing non-toxigenic A. flavus as a biological control agent and related products.

2.1.4 Variable-rate Application

Based on the previous studies, [7, 21] we have further conducted experiments to evaluate response of a variable-rate aerial application controller to changing flow rates and to improve its response at correspondingly varying system pressures[8]. The variable-rate application system consists of Differential Global Positioning System (DGPS) based guidance, an automatic flow controller, and a hydraulically controlled spray pump. The controller was evaluated for its ability to track desired flow rates set by the pilot and then the system was evaluated over several field trials to quantify its response to rapidly changing flow requirements and to determine the effect of the latest control algorithm improvements on response characteristics. The experiments illustrate an example of how iterative refinement of control algorithms in collaboration with the control system manufacturer could improve system response characteristics. System evaluation techniques described should also apply to aircraft that use propeller-driven spray pumps as well as hydraulically controlled spray pumps.

2.2 Low-Altitude Remote Sensing

With the research and development of aerial application technology scientists of USDA-ARS at Stoneville, Mississippi have also conducted studies of remote sensing, especially at low altitude, to provide site-specific information for aerial spray of crop protection and production materials. In the past two decades, scientists evaluated various multispectral and thermal cameras for crop field sensing on the Air Tractor 402B to fly over research farms. In recent years, small UAVs have been developed to carry portable cameras to sense the certain parts of the research farms with high resolution. Figure 2 shows the entire area (~300 ha) RGB image acquired on Air Tractor 402B and the color-infrared (CIR) image and the RGB image acquired on a small UAV for a field (~0. 3 ha) in the area. The resolution of the entire area image is about 50 cm per pixel with the flight altitude of 850 m while the resolution of the UAV images are about 2 cm per pixel with the flight altitude of 20 m.

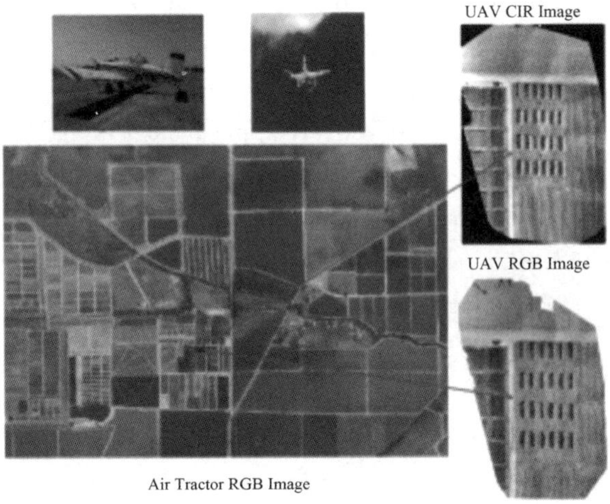

Figure 2 Entire area RGB image acquired on Air Tractor 402B and UAV CIR and RGB images of a field in the area for low-altitude remote sensing research in USDA-ARS CPSRU

With imagery acquired from the multispectral camera on Air Tractor 402B, an important study has been conducted to assess crop injury caused by off-target drift of aerially applied glyphosate. The widespread adoption of glyphosate-resistant (GR) crops in the U. S. has led to an unprecedented increase in glyphosate usage in recent years. Glyphosate is the most commonly applied herbicide either alone or with other herbicides to manage a broad spectrum of weeds. Pesticide drift, the physical movement of a pesticide particle onto an off-target, can occur when applied under weather conditions that promote drift. In virtually all pesticide applications, a small fraction of the pesticide drifts downwind and can be deposited on off-target surfaces. Off-target drift of aerially applied glyphosate can cause plant injury, which is of great concern to farmers and aerial applicators. To determine the extent of crop injury due to near-field drift, an experiment was conducted with a single aerial application of glyphosate. For identification of the drift effect on cotton,[19] corn and soybean plants, a field was planted in replicated blocks of cotton. Spray samplers were placed in the spray swath and in several downwind orientations to quantify relative concentration of applied chemical. An Air Tractor 402B spray airplane equipped with fifty-four CP-09 nozzles was flown down the center of the field to apply Roundup Weathermax and Rubidium Chloride tracer. Relative concentrations of this tracer were quantified at downwind spray samplers. At one, two and three week intervals aerial color-infrared imagery was obtained over the field using a GPS-triggered multispectral MS-4100 camera system. This study's main focus was to assess glyphosate spray drift injury to cotton using spray drift sampling and the color-infrared imagery. The processed drift and image data were highly correlated. The drift and image data were used as the indicators of percent visual injury in regressions with a strong ability of variability explanation. The results indicate that spray drift sampling and airborne multispectral remote sensing can be useful for determining crop injury caused by the drift of aerial applied glyphosate.

This study was extend to assessing crop injury caused by another herbicide, Dicamba, which is used for postemergence control of several broadleaf weeds in corn, grain sorghum, small grains, and non-cropland. For detection of crop herbicide injury, a field experiment was conducted in a soybean field at the research farm of the USDA-ARS CPSRU in Stoneville, Mississippi, USA.[23] A few weeks after dicamba treatment, RGB and CIR images were acquired with the cameras mounted on a small octocopter and a model fix-wing airplane flying over the field. This study indicated that the high-resolution UAV image data performed consistently well in characterizing image features within the crop yield in quantifying soybean injury from dicamba spray.

UAVs are a unique platform to perform crop field sensing with high resolution and least atmospheric interference without restriction of field conditions. The following are our other successful research activities of UAV remote sensing.

2.2.1 Build DSM to Estimate Plant Height

Low-altitude remote sensing with a small UAV was developed for estimation of crop plant height and hence estimation of the crop yield. The plant height was estimated by building the digital surface model (DSM) of the crop field by manipulating 3D point cloud data generated by stereo vision using the images acquired by UAV. The estimation of cotton and soybean plant height were highly accurate to infer the yield of the crops.[23]

2.2.2 Extract In-depth Image Features to Estimate Cotton Yield

This study showed that low-altitude remote sensing using a small UAV can offer reliable cotton yield estimation based on estimation of cotton unit coverage (CUC) by Laplacian image processing and identification of plots with poor illumination.[23] The relationship between estimated CUC and measured lint yield using the methods of direct image pixel intensity thresholding could not be well established while use of the Laplace operator to obtain the divergence of the gradient (spatial second derivative) of the image pixel intensity significantly improved the linear relationship of CUC with the lint yield after using the Laplacian method to extract the cotton boll features under canopy in the images.

2.2.3 Identifying Herbicide-resistant Weeds

As mentioned above, glyphosate is the most widely used herbicide, with increased frequency of use and amount in fields planted with GR crops. Repetitive and intensive use of glyphosate has exerted a high selection pressure on weed populations, resulting in the evolution of 37 GR weed species in the world (http://www.weedscience.org). Ten of them have appeared in Mississippi. Hyperspectral plant sensing techniques have been developed to effectively detect GR and GS Palmer amaranth (*Amaranthus palmeri* S. Wats.) and Italian ryegrass [*Lolium perenne* L. ssp. *multiflorum* (Lam.) Husnot] in greenhouse and soybean fields with detection rates of 90% and 80%, respectively.[24, 25] However, in-field hyperspectral plant sensing is still time-consuming and laborious because the current sensors are either operated on a slow-moving tractor for imaging certain areas in the field, or handheld by a technician to measure canopy spectra at certain points in the field. This tedious manner of hyperspectral data acquisition is an obstacle to extend the research results to practical uses. Use of UAV is an innovative way to fly over a crop field to rapidly determine the distribution of weeds. We have been undertaking a research project to mount a portable hyperspectral sensor on a small UAV to overfly a soybean field at a very low altitude to quickly determine the distribution of GR and GS weeds.

2.2.4 Field Observation Scale Optimization and Multisource Coordination

Currently crop monitoring remote sensing analysis is mostly based on remote sensing data in a single spatial scale. However, in practice low spatial resolution or high spatial resolution remote sensing data are all limited to meet all requirements of crop growth studies and crop production management. Therefore, multi-scale remote sensing for crop monitoring has been more and more studied and applied to improve single-scale remote sensing. We have been conducting studies since 2015 to use digital RGB and multispectral cameras on multirotor UAVs to fly over the research farms of USDA-ARS, CPSRU at Stoneville, Mississippi, USA. The flyovers were conducted at different crop growth stages and varied flight altitudes. At the same time high-resolution satellite imagery were analyzed with the data acquired from the ground-based systems to optimize the scale of field observation for crop growth status and identify crop stress caused by multiple factors.

3 Agricultural Aviation and Precision Agriculture in the Next Decade

In the next phase of our research projects we will focus on integration of application technology and remote sensing systems to improve precision operations for crop protection and production. Processing and analysis of massive or big data will be our focus for decision support in precision agriculture. Remote sensing big data will be our emphasis to study the theory and practice of data science for precision agriculture. Agricultural remote sensing data, as general remote sensing data, have all characteristics of big data with its vast volume and complexity by generating earth-observation data and analysis results daily from the platforms of satellites, manned/unmanned aircrafts, and ground-based systems. The acquisition, processing, storage, analysis and visualization of agricultural remote sensing big data are critical to the success of precision agriculture. We are investigating the theory and practice of agricultural remote sensing big data management for data processing and applications. A well-established five-layer-fifteen-level satellite remote sensing data management structure[26, 27] is adopted to create a more appropriate three-layer-nine-level remote sensing data management structure for management and applications of agricultural remote sensing big data for precision agriculture where the sensors are typically on high-resolution satellites, manned aircrafts, UAVs and ground-based systems.

In the next ten years or so, agricultural aviation will be advanced greatly and hence precision agriculture. Agricultural airplanes will be developed with high capacity of chemical loading up to 3,800 liters (1,000 gallons) from the current capacity from 340 to 3,000 liters (90 ~ 800 gallons). UAV-based plant protection will be greatly developed and widely applied, especially in China. However, in U. S. , UAV spraying systems will stay in research and development and won't be much used commercially due to powerful manned aerial and ground-based systems effectively operating in the country. In-flight (between-flight) control of aerial application will be renovated with desired droplet spectrum regardless of the composition of the chemical mixture or environmental conditions. Aerial variable-rate technology will be ready for practice.

Information systems, remote sensing technology and data science will be integrated to improve the performance of precision agriculture. Agriculture will be more considered under ecological system and precision crop farming will be more closely connected to precision livestock/animal farming.

Innovated methods, optimized algorithms, massive data, and super computing power will be ready for smart/intelligent agriculture in the cyber-physical architecture. Agricultural automation will dominate agricultural operations with advanced materials, mechanical and electronics science and technology. Ubiquitous collaborative robots will be on farmers' hands. Artificial intelligence with deep learning will be developed to advance plant phenotyping and perception of agricultural robotic systems. Real-time agricultural data processing, analysis, control and adaptation will take important roles for decision support in agricultural management.

References

[1] LAN Y, THOMSON S J, HUANG Y, et al. Current status and future directions of precision aerial application for site-specific crop management in the USA [J]. Computers and electronics in agriculture, 2010, 74: 34 -38.

[2] ZHANG C, KOVACS J M. The application of small unmanned aerial systems for precision agriculture: a review [J]. Precision agriculture, 2012, 13 (6): 693 -712.

[3] SATO A. The RMAX helicopter UAV, public report, aeronautic operations [M].

Shizuoka: Yamaha Motor Co., Ltd., 2003.

[4] ZHOU Z, ZANG Y, LU X, et al. Technology innovation development strategy on agricultural aviation industry for plant protection in China [J]. Transactions of the CSAE, 2013, 29 (24): 1-10.

[5] KELEMEN D. The future of unmanned aircraft systems: is there a niche in aerial application? [J] Ag. aviation, 2013, 9 (10): 15-21.

[6] GRISSO R, ALLEY M, THOMSON W, et al. Precision farming tools: variable-rate application [M]. Virginia Cooperative Extension Publication, 2011: 442-505.

[7] THOMSON S J, SMITH L A, HANKS J E. Evaluation of application accuracy and performance of a hydraulically operated variable-rate aerial application system [J]. Transactions of the ASABE, 2009, 52 (3): 715-722.

[8] THOMSON S J, HUANG Y, HANKS J E, et al. Improving flow response of a variable-rate aerial application system by interactive refinement [J]. Computers and electronics in agriculture, 2010, 73: 99-104.

[9] YANG C, MARTIN D E. Integration of aerial imaging and variable-rate technology for site-specific aerial herbicide application [J]. Transactions of the ASABE, 2017, 60: 635-644.

[10] HUANG Y, THOMSON S J. Characterization of in-swath spray deposition for CP-11TT flat-fan nozzles used in low volume aerial application of crop production and protection materials [J]. Transactions of the ASABE, 2011, 54 (6): 1973-1979.

[11] HUANG Y, THOMSON S J. Characterization of spray deposition and drift from a low drift nozzle for aerial application at different application altitudes [J]. International journal of agricultural and biological engineering, 2011, 4 (4): 1-6.

[12] REDDY K N, DING W, ZABLOTOWICZ R M, et al. Biological responses to glyphosate drift from aerial application in non-glyphosate-resistant corn [J]. Pest management science, 2010, 66: 1148-1154.

[13] HUANG Y, THOMSON S J, ORTIZ B V, et al. Airborne remote sensing assessment of the damage to cotton caused by spray drift from aerially applied glyphosate through spray deposition measurements [J]. Biosystems engineering, 2010, 107: 212-220.

[14] HUANG Y, DING W, THOMSON S J, et al. Assessing crop injury caused by aerially applied glyphosate drift using spray sampling [J]. Transactions of the ASABE, 2012, 55 (3): 725-731.

[15] THOMSON S J, HUANG Y, FRITZ B K. Atmospheric stability intervals influencing the potential for off-target movement of spray in aerial application [J]. International journal of agricultural science and technology, 2017, 5 (1): 1-17.

[16] HUANG Y, THOMSON S J. Atmospheric stability determination at different time intervals for determination of aerial application timing [J]. Journal of biosystems engineering,

2016, 41 (4): 337-341.

[17] TESKE M E, THISTLE H W, ICE G G. Technical advances in modeling aerially applied sprays [J]. Transactions of the ASABE, 2003, 46 (4): 985-996.

[18] TAGUCHI G. System of experimental design [M]. Dearborn MI: Unipub/Kraus/American Supplier Institute, 1987.

[19] HUANG Y, ZHAN W, FRITZ B K, et al. Analysis of impact of various factors on downwind deposition using a simulation method [J]. Journal of ASTM international, 2010, 7 (6): 1-11.

[20] HUANG Y, ZHAN W, FRITZ B K, et al. Optimizing selection of controllable variables to minimize downwind drift from aerially applied sprays [J]. Applied engineering in agriculture, 2012, 28 (3): 307-314.

[21] SMITH L A. Automatic flow control for aerial applications [J]. Applied engineering in agriculture, 2001, 17 (4): 449-455.

[22] HUANG Y, THOMSON S J. BRAND H J, et al. Development and evaluation of low-altitude remote sensing systems for crop production management [J]. International journal of agricultural and biological engineering, 2016, 9 (4): 1-11.

[23] HUANG Y, BRAND H J, SUI R, et al. Cotton yield estimation using very high-resolution digital images acquired with a low-cost small unmanned aerial vehicle [J]. Transactions of the ASABE, 2016, 59 (6): 1563-1574.

[24] REDDY K N, HUANG Y, LEE M A, et al. Glyphosate-resistant and glyphosate-susceptible Palmer amaranth (*Amaranthus palmeri* S Wats.): hyperspectral reflectance properties of plants and potential for classification [J]. Pest management science, 2014, 70: 1910-1917.

[25] LEE M A, HUANG Y, NANDULA V K, et al. Differentiating glyphosate-resistant and glyphosate-sensitive Italian ryegrass using hyperspectral imagery [J/OL]. Proc. SPIE 9108, sensing for agriculture and food quality and safety Ⅵ, 91080B, 2014. DOI: 10.1117/12.2053072.

[26] YU T, LI J, CHENG T, et al. Data grading organization method based on longitude and latitude grid: CN102346923 A [P]. Patent of P. R. China. No. 2012.

[27] Gu X, Yu T, XIE D, et al. A hierarchically organized method based on the latitude and longitude grid data: CN102346923 B [P]. 2013.

作者简介

Yanbo Huang（黄岩波），博士，现任美国农业部农业研究服务局作物生产系统研究所高级研究员，得克萨斯农工大学、密西西比州立大学和三角洲州立大学的兼职教授，同时担任美国农业部小企业创新研究基金工程技术专业评审委员会主任，美国农业与生物工程师协会海外华人农业、食品与生物系统工程师协会（AOC）主席及 *International Journal of Agricultural Science and Technology* 期刊主编和多个国际期刊的副主编。长期从事农业信息化、过程建模与控制等方面的研究工作。发表学术论文 160 余篇，主编了 *Automation for Food Engineering: Food Quality Quantization and Process Control*。

农用无人机在新疆现代农业中的应用

赵 岩 陈学庚 温浩军①

1 新疆绿洲农业的主要特点

新疆农业属于典型的绿洲农业，拥有大小绿洲8000多片，呈“岛屿”式分布在天山南北干旱荒漠中，绿洲多为河流冲积而成的冲积扇绿洲平原或是河流流域开发形成的平原。[1]新疆经济作物的播种面积比例明显高于全国，初步形成了以棉花、林果、畜牧和优质粮食为主的农业格局。[2]新疆粮食作物以小麦、玉米、水稻为主，[3]经济作物中棉花占主导位置，棉花种植也是新疆地区的支柱产业[4]。

新疆特殊的光热和气候资源适合多种农作物生长，利于粮食作物高产。据报道，2013年，奇台农场3.6公顷小麦平均亩产734.7千克；2014年，新疆兵团第四师88团80亩小麦平均亩产753.96千克。2013年，奇台农场连片种植833公顷玉米平均亩产1126.4千克；2015年，小面积试验田平均亩产1511.74千克。2013—2014年，兵团第一师一团100亩水稻旱直播平均亩产超过1000千克。据国家统计局数据，2016年全国棉花总产量529.9万吨，新疆359.4万吨，新疆棉花产量占全国总产量的67.82%；全国棉花种植总面积3344.74千公顷，新疆1805.15千公顷，新疆棉花种植面积占全国棉花种植总面积的53.97%。2014年，兵团皮棉亩单产达155.7千克，比全国单产高出58.12千克。新疆棉花总产、单产、品质和调出量均居全国首位。[5]

新疆是我国特色林果业生产区，是世界六大果品生产带之一，资源优势、区位优势明显，发展特色林果业条件得天独厚。南疆与世界最著名的果品产地美国加利福尼亚州的生态环境基本相似，北疆葡萄主产区与法国波尔多葡萄产地处在相近纬度，生产的特色果品单产高、品质好。新疆已基本形成南疆环塔里木盆地以红枣、核桃、巴旦木、杏、香梨、苹果为主栽树种的特色林果主产区，东疆吐哈盆地以鲜食葡萄、红枣为主的优质特色林果基地，北疆伊犁河谷和天山北坡以鲜食和酿酒葡萄、枸杞、小浆果、时令水果为主的特色鲜明的林果基地。[6]新疆是我国第一大加工番茄产区、第一大葡萄产区、第一大杏产区、库尔勒香梨的唯一产区、第二大红枣产区、唯一大面积种植巴旦木和开心果的产区。新疆地区番茄酱出口量占国际贸易的四分之一，产量居全国首位。啤

① 赵岩、陈学庚、温浩军，石河子大学机械电气工程学院，农业农村部西北农业装备重点实验室。

酒花、枸杞、红花、打瓜产量占全国产量的比例很大，葡萄、香梨、哈密瓜、红枣和苹果等享誉全国。

新疆现代节水灌溉设施农业在我国节水农业中占有十分重要的地位，已成为全国最大的高效节水示范区。[7] 水资源匮乏是新疆现代农业发展面临的核心问题，在西北农业灌溉用水效率研究中，新疆具有很强的代表性。新疆是我国西部干旱地区，水资源利用总量中农业灌溉用水占90%以上，提高农业用水效率成为应对新疆农业水资源短缺的有效途径。大田棉花膜下滴灌是新疆高效节水的主导技术。与传统灌溉方式相比，膜下滴灌在节水、节肥、节地、节劳、增产和增收方面效果显著。新疆棉花膜下滴灌技术在我国大田作物种植中的大规模应用，被誉为当前农业技术的一次飞跃。滴灌节水技术是以色列发明的，主要用于种植高附加值经济作物，器材的价格昂贵，在中国无法大面积推广使用。新疆在滴灌节水方面的重大贡献是实现低成本器材和装备的研究开发，并形成服务体系，大面积应用于大田作物。2017 年年末，新疆有效灌溉面积为 125.8 万公顷，其中，高新节水灌溉面积 104.2 万公顷。

先进农业技术在新疆地区应用广泛。新疆耕地多为平坦土地，耕地条田规整、单块面积大，人均种植规模大。农机与农艺相结合，适合规模化农机作业，有利于抢农时，保证适时作业，而且作业质量也能得到明显提升[8]，各类先进农业技术得到推广应用[9]。新疆兵团是新疆先进农业技术的示范区。《新疆生产建设兵团 2017 年国民经济和社会发展统计公报》数据表明，新疆兵团 2017 年种植业耕种收综合机械化率已达 94%，机采棉面积 53 万公顷（795 万亩），棉花机采率 80%，畜牧业机械化水平较高。目前，新疆兵团小麦、水稻、玉米等主要粮食作物已实现全程机械化作业，耕深监测、卫星导航拖拉机自动驾驶、无人机遥测和植保等先进农业技术已得到示范应用。新疆棉花全程机械化技术体系初步建成，并在山东、河北等棉区进行技术引进和试验示范。

2　农用无人机在新疆农业中的应用现状

新疆耕地较多呈“条田”形状，便于集约化管理，适合农业植保无人机作业，其应用和推广对新疆农业现代化发展起到了推动作用。近年来，农用无人机技术在新疆昌吉、奇台、哈密、和田等地区进行了推广应用，主要用于航空植保、灾情检测和地形测绘等作业。[10-12] 昌吉州是无人机技术应用示范较为成功的地区之一。昌吉拥有耕地约 53 万公顷，是全国的商品粮、商品棉、制酱番茄、酿酒葡萄生产基地。2014 年，昌吉开始建设农业航空产业，通过推行适应新疆特点的农用无人机推广新机制、新模式，初步形成了“政府引导、企业领办、部门联动、企业化经营、科技创新驱动、产学研用结合”的发展格局，通过大面积、多品种的飞防示范作业，让农户接收和认可。2015—2017 年，昌吉州累计完成作业面积 120 千公顷。同时，将无人机技术拓展和延伸到小麦、棉花、瓜菜大田作物、高秆作物和山地作物的病虫害防治、森林有害生物防治、草原灭蝗灭鼠等多个领域。昌吉地区无人机技术的发展得到了中国工程院院士罗锡文教授的肯定，他称之为“昌吉模式”。综合分析，其成功的主要因素，一是由政府引导，整合各种资源，实现多部门协同推进；二是构建州、县、乡飞防网络服务体系和跨地区连锁应用体系；三是结合新疆实际，以合作社、家庭农场和种植大户为载体，建立示范区，采用多种方式服务农户；四是建立产、学、研平台和试验示范基地，引进前沿技术和创新成果，不断拓宽应用领域。

3 农用无人机在新疆农业中的应用趋势

与传统技术相比，农用无人机使用方便、快捷，目标针对性强，获取数据的速度快、精度高，应用优势明显，现代农业对农业航空的依赖性将更加迫切。[13-17]新疆粮食和棉花种植超过420万公顷，农用无人机能承担新疆20%的防控面积，经济社会效益十分显著。无人机技术可以在新疆农用中应用的主要方向包括土壤信息遥测、棉花长势遥测、脱叶剂喷施、作物产量估测、滴灌旱情监测、病虫害灾情防治、林果植保和授粉、防护林带航空植保等农业生产环节。

用于土壤信息遥测。无人机遥测系统可实现影像获取和数据处理全过程自动化，能科学准确地评估土壤养分含量，对后期进行种植区域优化布局、土壤改性、精准施肥等作业具有指导意义。目前，无人机遥测系统主要的手段是卫星遥感监测，实时性和精准度都存在一定局限，利用无人机与云数据服务平台相结合的方式可极大地提高信息的实时性和可靠性，科学合理地指导相关技术作业。[18]

用于脱叶剂喷施。新疆的棉花生产已初步实现全程机械化。2015年，新疆地方和兵团棉花机采面积超过72万公顷。全面实现棉花生产全程机械化是奋斗目标，提高机采棉品质是攻关方向，脱叶剂喷洒质量的提高是关键手段。大型农用飞机曾用于大面积喷洒脱叶剂，但由于林带等环境影响，两端飞行高度较高，地头喷洒效果不佳。无人机喷洒脱叶剂全面达到质量要求，需求量将非常大。[19]

用于作物产量估测。收获前进行作物产量评估，对调配采收机械、仓储物流资源以及加工资源有重要意义。[20]目前，新疆地区作物产量评估主要依靠人工进行，效率低，劳动强度大。依靠无人机遥感方法进行产量评估，可以快速、准确估测作物产量，对推进农业信息化和提升农业现代化水平具有重要意义。[21]

用于喷洒农药进行病虫害灾情防治。对地面植保机械不适合承担的防控工作，无人机可进行有效补充，使农作物病虫害防治从地面防治变为空中防治。[22-25]采用无人机喷洒可节省劳动力，人员安全得到保障，能节约30%的农药使用量，喷洒农药质量达标，价格适中，并可大幅度降低劳动力成本，符合农业农村部“一控两减三基本”任务要求。目前，新疆棉花种植面积200万公顷左右，植保和病虫害防治主要依靠地面施药机械和人力辅助进行。棉花病虫害的发生初期一般是呈点、片状，其传播途径一般是扩散，采用拖拉机或人工进地喷洒农药容易导致病虫害扩散速度加快。利用无人机监测发现病虫害，进行点片防控是最佳选择。将病虫害测报信息链中的相关技术进行推广应用，将极大地提升病虫害测报部门的技术水平和工作效率。提前预测和防治季节性或爆发性病虫害，能把农产品的损失降到最低。

新疆农耕土地面积大，林果业和畜牧业的绿色发展特色鲜明，农用无人机技术在新疆农业发展中有较为广阔的空间，也将为提升新疆农业现代化水平起到重要支撑作用。

4 对农用无人机在新疆地区示范推广的建议

新疆农业主要是平作，地面行走植保机械发展比较完善。无人机是近年来发展起来的新技术，优势明显，但不能完全替代地面植保机械，它们优势互补，地面行走植保机械与无人机植保存在相当长的共存期。根据新疆农业的特点，提出农用无人机技术在新疆地区示范推广的建议如下。

选择载重量合适的无人机。载重量小、续航时间短是无人机目前存在的普遍问题。载药量 15 ～ 50 千克符合新疆实际情况，特别是新疆兵团，土地规模较大，起飞 1 次要完成较大作业面积，从性价比方面讲，15 ～ 50 千克符合多数情况的要求。载重量大当然好，但价格也明显上升。[26]

加快农用无人机标准化体系建设。农用无人机领域的标准化工作起步较晚，产品规范没有形成完整的标准体系，已颁布实施的大多以通用要求为主，具体涉及农用无人机的标准和安全操作规程则不完善，缺乏针对性。农用无人机标准化体系建设是当前迫切需要解决的问题，应尽快制定地方或者行业标准以及安全操作规程。农用无人机归属管辖也是迫切需要确定的问题。[27]

提升改进现有无人机技术。[28-30] 使用可靠、使用安全、使用经济是无人机健康发展的关键；皮实耐用，抗高温、高湿、灰尘和农药腐蚀是无人机的普遍要求。农用无人机操控水平是人们关注的目标，操控要逐步实现智能化，要具备定位记忆功能，在农药喷洒作业中，接行要准，要能自动找准上次结束位置，保证不重不漏。低空遥感是农情信息获取的配套设施，适用于无人机遥感的传感器、农情遥感信息获取相机、农情信息提取分析软件属薄弱环节，需继续加强攻关。无人机专用喷药喷头的研制是攻关目标，喷药机械装备要与农药特性相融合。

加强无人机技术服务体系建设。[31] 地方政府通过政策支持，鼓励基层农业合作社或社会团体组建无人机作业公司，在区域内做好农户服务，收取合适的作业费用。考虑到成本投入和行业风险，我们不主张农户单独购买无人机。

参考文献

［1］樊自立，艾里西尔，王亚俊，等. 新疆人工灌溉绿洲的形成和发展演变［J］. 干旱区研究，2006（6）：410－418.

［2］孟梅. 新疆农业产业结构调整与农地资源优化配置研究［D］. 乌鲁木齐：新疆农业大学，2014.

［3］吐尔逊·买买提，谢建华. 新疆农业机械化发展水平区划时空格局［J］. 中国农业资源与区划，2017，（2）：81－88.

［4］史利洁. 西北旱区粮食作物生产水足迹空间差异分析［D］. 咸阳：西北农林科技大学，2016.

［5］瞿建蓉. 新疆绿洲生态农业发展思考［J］. 实事求是，2017，（5）：74－77.

［6］孙兰凤，安尼瓦尔·阿木提. 可持续视角下的新疆特色林果业发展研究

[D]. 乌鲁木齐：新疆大学，2009.

[7] 林海，胡锡宁，肖光顺，等. 新疆精准农业的发展策略 [J]. 新疆农业大学学报，2002，(3)：83 - 86.

[8] 陈学庚，赵岩. 新疆兵团农业机械化现状与发展趋势 [J]. 华东交通大学学报，2015，(2)：1 - 7.

[9] 江雪丽. 新疆农业现代化发展水平分析与评价 [J]. 山东纺织经济，2017，(2)：68 - 71.

[10] 吴兵. 新疆农业发展问题探析 [J]. 科技经济市场，2015，(8)：29.

[11] 韩琳. "一带一路" 背景下新疆农业水土资源配置研究 [J]. 中国农业资源与区划，2017，(9)：115 - 121.

[12] 王应宽. 农用无人机为现代农业插上高科技的翅膀 [J]. 农业工程技术，2018，38 (9)：6 - 8.

[13] 张忠喜. 无人机农药喷洒系统喷雾特性及影响因素探究 [J]. 石河子科技，2018，(2)：8 - 10.

[14] 李晨辉. 谈 YR-GSF06 型植保无人机在昌吉州农业生产中的运用 [J]. 农机使用与维修，2015，(5)：102 - 103.

[15] 兰玉彬，王国宾. 中国植保无人机的行业发展概况和发展前景 [J]. 农业工程技术，2018，38 (9)：17 - 27.

[16] 娄尚易，薛新宇，顾伟，等. 农用植保无人机的研究现状及趋势 [J]. 农机化研究，2017，39 (12)：1 - 6，31.

[17] 巩春源，刘铁军. 无人机农业植保的现状分析和产业发展 [J]. 农业开发与装备，2016，(09)：136，150.

[18] 贾鹏宇，冯江，于立宝，等. 小型无人机在农情监测中的应用研究 [J]. 农机化研究，2015，37 (4)：261 - 264.

[19] 杲先民，任燕成，刘涛荣. 大疆植保无人机喷施棉花脱叶剂试验 [J]. 农村科技，2017，(5)：28 - 29.

[20] 汪沛，罗锡文，周志艳，等. 基于微小型无人机的遥感信息获取关键技术综述 [J]. 农业工程学报，2014，30 (18)：1 - 12.

[21] 王利民，刘佳，杨玲波，等. 基于无人机影像的农情遥感监测应用 [J]. 农业工程学报，2013，29 (18)：136 - 145.

[22] 张东彦，兰玉彬，陈立平，等. 中国农业航空施药技术研究进展与展望 [J]. 农业机械学报，2014，45 (10)：53 - 59.

[23] 姜锐，周志艳，徐岩，等. 植保无人机药箱液量监测装置的设计与试验 [J]. 农业工程学报，2017，33 (12)：107 - 115.

[24] 单磊. 无人机系统综合标准化思考 [J]. 中国标准化，2015，(6)：108 - 113.

[25] 巩春源，刘铁军. 无人机农业植保的现状分析和产业发展 [J]. 农业开发与

装备，2016，(9)：136，150.

［26］温源，薛新宇，邱白晶，等. 中国植保无人机发展技术路线及行业趋势探析［J］. 中国植保导刊，2014，34（S1）：30－32.

［27］蒙艳华，周国强，吴春波，等. 我国农用植保无人机的应用与推广探讨［J］. 中国植保导刊，2014，34（S1）：33－39.

［28］王术波，陈建，彭兵忠. 我国农用无人机产业链分析［J］. 中国农业大学学报，2018，23（3）：131－139.

［29］周志艳，明锐，臧禹，等. 中国农业航空发展现状及对策建议［J］. 农业工程学报，2017，33（20）：1－13.

［30］伊张芸. 农用无人机的发展现状与思考［J］. 现代农机，2017，(5)：14－16.

［31］冷志杰，蒋天宇，刘飞，等. 植保无人机的农业服务公司推广模式研究［J］. 农机化研究，2017，39（1）：6－9.

作者简介

赵岩，硕士，副研究员，河南省永城市人。现任农业农村部西北农业装备重点实验室办公室主任。主要从事棉花生产全程机械化和残膜回收技术装备研究，研究内容包括残膜回收技术装备研发与应用、国产棉花收获和清理加工技术装备的应用示范、棉花收获机械化、精准对行分层施肥技术及装备研究示范等。成果获省部级科技奖励一等奖1项、二等奖1项、三等奖3项，获国家专利30余项，发表论文30余篇，参编行业著作2部，参编地方标准1项。

陈学庚，研究员，农业机械设计制造专家，江苏省泰兴市人。中国工程院院士，现任石河子大学博士生导师，中国农业机械学会名誉理事长，农业农村部西北农业装备重点实验室主任。扎根边疆基层一线从事农机研究和推广工作54年，主要从事旱田作业机械的研究。为新疆棉花生产全程机械化技术体系的建设和大面积推广应用作出了重大贡献，突破了地膜植棉机械化关键技术，攻克了滴灌技术大规模应用农机装备难题，研发了棉花生产全程机械化关键技术与机具，为促成新疆棉花产量两次飞跃提供了有力的农机装备支撑。扎根新疆，建设创新团队，培养高级科技人才40余名。获省部级以上科技奖励23项，其中，国家科技进步一等奖1项、二等奖2项，省部级一等奖7项；获国家专利120余项，其中，发明专利58项，专利技术转化后获国家优秀新产品9项；发表论文100余篇，撰写专著4部。获“全国

杰出专业技术人才”、中华农业英才奖等荣誉称号和奖项16项。

温浩军，研究员，农业机械设计制造专家，甘肃临洮人。现任农业农村部西北农业装备重点实验室副主任，新疆农业机械学会副秘书长。长期在基层一线从事农业机械研究、开发和推广工作。在棉花生产全程机械化关键技术及装备的研发应用开发出多种新产品，在农业生产中广泛应用；在化肥农药减施方面开发了大型自走式智能化高效喷雾机，已经在新疆、东北开始应用；在农田残膜污染治理工作方面，负责开发的秸秆还田与残膜回收联合作业机残膜回收率达90%以上，回收残膜含杂率低，为残膜的资源化利用创造了条件。先后获国家科技进步二等奖2项，省部级奖励16项。发表论文80余篇，获国家专利88项，先后获中国青年科技奖、全国五一劳动奖章、新疆兵团科学技术突出贡献奖等荣誉称号。

Drift Management Approaches in Precision Aerial Pesticide Application: A Review

Andrew John Hewitt① **Juan Liao**②

1 Introduction

The use of pesticide continues to be the most important approach to control pests, weeds, diseases and pathogens for crops.[1] Recently, with the increase of pesticide using in plant protection, the concerns over the influence of pesticide drift on environment and public health increased as well.[2, 3] Precision spraying requires that the pesticides must be sprayed on the target areas with the correct liquid amount and minimal drift. Inversely, imprecision spraying could result in economic losses, damage to environment (such as water, soil, air, etc.), exposing operators and livestock in danger, residents to agricultural products, and phytotoxicity to terrestrial and aquatic ecosystems.[4, 5] However, during the spraying progress, a part of liquid, which was reported to be as much as 50% of the total spraying liquid amount[6], would drift to the undesirable areas instead of depositing on the intended areas, leading to a lower dose than intended on the target area. As an important measure for saving economic resources, ensuring the safety of environment, human beings and other creatures, the development of precision application has significant importance for pesticide spraying.[7, 8]

Aerial spraying[9, 10] has been defined as one of the most effective spraying methods currently. Aerial spraying can spray fast, with facilitating the timing requirement under optimal conditions such as ideal meteorology, and ability to deal with unpredictable emergency operation tasks. However, their greater height and vortices than ground spraying present challenges to reaching only the intended target at the optimal rate, and a large part of liquid may drift to undesired areas. Therefore, researchers have been focusing on technologies which could promote the development of precision aerial pesticide application (PAPA) and reduce spray drift for several decades. There are many approaches which can improve the ability of

① Andrew John Hewitt, Centre for Pesticide Application and Safety, University of Queensland.

② Juan Liao, College of Engineering, South China Agricultural University/National Center for International Collaboration Research on Precision Agricultural Aviation Pesticides Spraying Technology, Key Laboratory of Key Technology on Agricultural Machine and Equipment, Ministry of Education.

PAPA, such as new types of nozzles, variable rate (VR) spraying[11], soil and crop nutrient mapping[11, 12], pesticide dosage adjustment according to crop environment (PACE)[13, 14], global navigation satellite system (GNSS), geographic information system (GIS)[15], and so on. Recently, the development of PAPA has been promoted by the development of the above technologies. However, extensive efforts are still needed to be carried on to promote the development of PAPA further.

Reduction of off-target spray caused by aerial application of pesticides has been recognized as a primary necessity for both cost-effective application and environmental protection. However, in the application of pesticide spraying, a complex interaction of variables is involved. The drift reduction technologies (DRTs) include equipment design and configuration, meteorological factors, application parameters, tank mix characteristics, and crop canopy effects. Although great effort has been made for drift management all over the world, spray drift is still a serious problem in agricultural spraying. Nozzle, as the most important component of aerial and ground spraying, can decide the droplet size spectrum in spraying, and its development significantly influences precision spraying. Numerous types of nozzles, including air induction nozzles, electrostatic nozzles, have been developed for drift reduction in PAPA. Meanwhile, prediction and calculating models have been developed to help drift reduction in spraying. This article aims to reveal the current development status and analyze the future work of DRTs in PAPA through the development of new nozzle types, models and adjuvants to provide guidance for facilitating the development of PAPA.

2 New Types of Nozzles

Because nozzles have the ability to provide a wide range of droplet size, they were developed liquid atomizing in agricultural spraying. But most traditional nozzles do not take good control of spraying drift. With the purpose of improving the precision spraying, new types of nozzles, including air induction nozzles (AINs)[16], electrostatic nozzles, multiple orifice solid stream nozzles, spinning disc nozzles, slotted and porous drum nozzles had been developed to narrow the relative span (RS) of droplets for the uniform of spray droplets.

2.1 Larger Droplet Production Nozzles

Coarse droplet spectra sprays drift less than fine droplet spectra sprays, but the drift reduction conflicts with application efficacy. Air induction nozzles (AIN), Accu-Flo nozzles[17], hollow cone nozzles, Beecomist atomizer[18], Micronair nozzles[19], Thru Valve boom (TVB), CP nozzles, etc. (see Fig. 1) that produce a narrow range of droplet sizes (small value of RS) are desirable for drift control because fewer fine droplets are produced for a given medium droplet diameter.

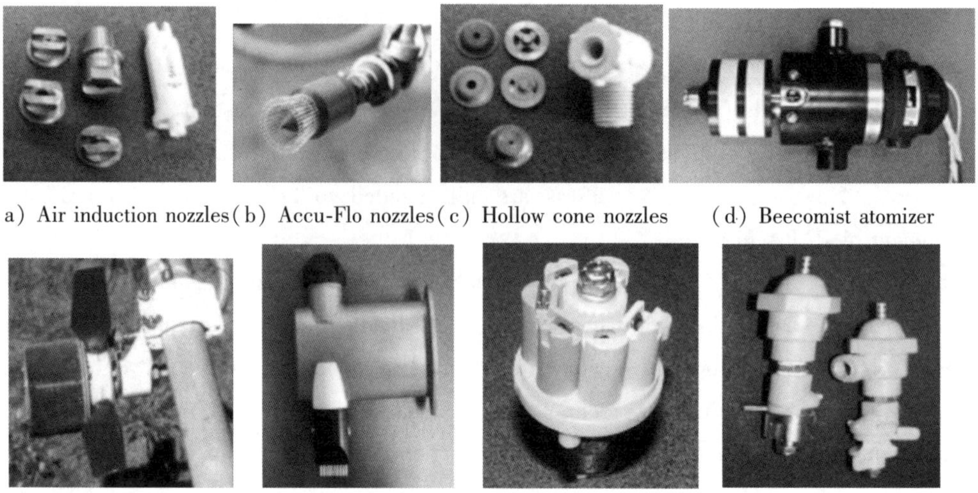

(a) Air induction nozzles (b) Accu-Flo nozzles (c) Hollow cone nozzles (d) Beecomist atomizer

(e) Micronair nozzles (f) Thru Valve boom (g) CP nozzles

Figure 1 The new types of nozzles for drift reduction in PAPA

AIN discharges large bubble-containing droplets which reduce the number of small droplets and creates air bubble in the larger droplets to aid their dispersion.[20, 21] And the droplets explode on impact with the target to produce small droplets to ensure greater coverage than conventional sprays. Accu-Flo nozzle (Bishop Equipment Co., Hatfield PA, USA) applies droplets uniformly through several needles in a radial pattern. And TVB nozzle (Waldrum Specialties, Southhampton, PA, USA) was developed by using the similar principle as Accu-Flo nozzle, the Accu-Flo nozzle and TVB nozzle shatter droplet differently from the traditional nozzles.

Swirl-type hollow-nozzles based on fluid inertia to transmit fluid to the outside of the nozzle, overcome surface tension and produce a hollow cone, thereby enhance the break-up and atomization process.[23] Beecomist atomizer could be quickly adapted to a wide variety of aircrafts for ULA spraying and has the advantages of dispensing formulated chemicals at extremely low volumes.[24] Minogure[22] observed a very low percentage of droplets (0.2%) smaller than 153 μm diameter (which is prone to drift) from the nozzles.

2.2 Electrostatic Nozzles

Electrostatic nozzles[25, 26] [see Fig. 2 (a)] were developed to increase the chemical deposition on crops and improve the pesticide bio-efficacy and efficiency via reducing off-target pesticide into the environment. A high voltage is applied to the liquid in order to cause the liquid at the atomization tip to break up into a population of electrostatically charged droplets, which has been observed in some field studies to produce a 1.6- to 24-fold increase in spray deposition over conventional spraying application methods.[27] Electrostatic nozzles are fitted on aircrafts [see Fig. 2 (b)] to help the droplets to penetrate and deposit on crop targets and reduce the chemical drift into the environment.

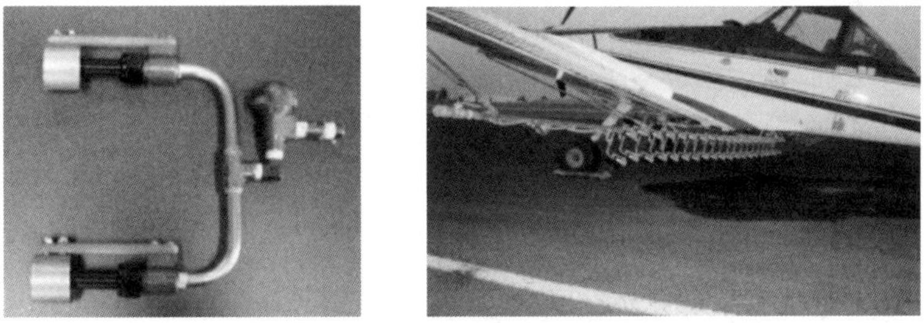

(a) Electrostatic nozzle (b) Electrostatic nozzles fitted on aircraft

Figure 2 Electrostatic nozzle and electrostatic nozzles fitted on aircraft

From the droplet size and RS comparison results of different nozzles showing in Fig. 3 and Fig. 4, the RS range of most traditional hydraulic nozzles, rotary cage atomizers, electrostatic nozzles, multiple orifice stream solid nozzles and spinning disc nozzles are: 1.2 – 1.6, 1.2 – 1.6, 0.8 – 1.3, 0.4 – 0.8 and 0.4 – 0.5 respectively. It was proved that the new types of nozzles could provide narrow ranges of RS, which is good for pesticide application efficacy. It also demonstrated that the new types of nozzles could improve pesticide application efficacy and reduce the chemical drift.

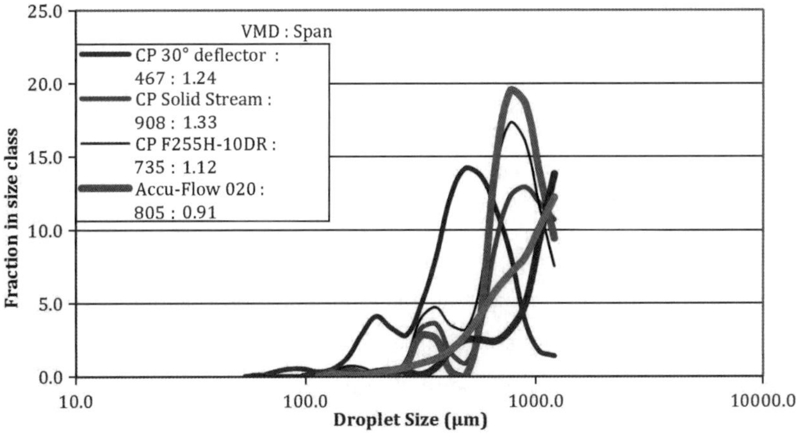

Figure 3 Droplet size and RS comparison of rotary nozzles

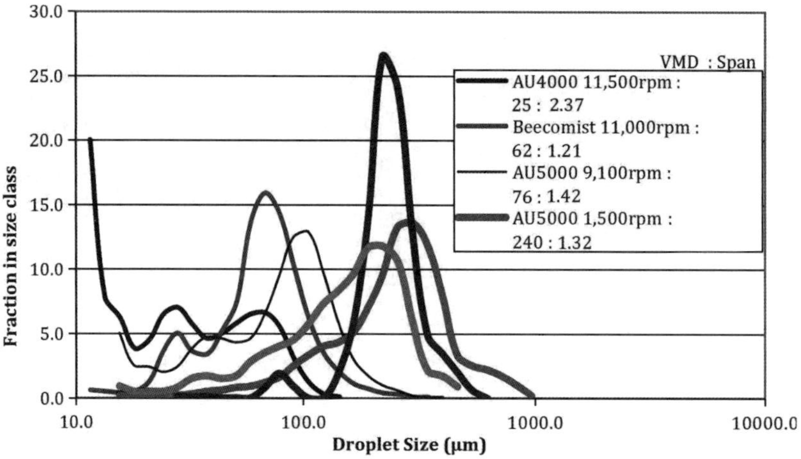

Figure 4 Droplet size comparison of different types of nozzles

3 Models

In aerial pesticide spraying, good efficacy requires good coverage, which is affected by spray dynamics, flight speed, physical (such as surface tension dynamics) and chemical properties of tank mix, target crops characteristics [the leaf area index(LAI), crop density, etc.], weather, etc. But in the field study, the issue is that the system which works well at one spraying site may not work so well at another because of the conditions variety. When we change the chemistry formulation we can change not only droplet size but also the spreading and uptake, and formulation effects can vary widely with different nozzles, pressures, etc. Since it is not easy for user to make a decision for a spraying task to a certain circumstance, models and mathematic simulations are very important tools to help with simplifying spray tasks.[28] Different types of models, including atmospheric dispersion models, droplet trajectory models, and statistical models have been used in agricultural spray.

3.1 Droplet Trajectory Models

Droplet trajectory models were developed for determining the movements and positions of individual droplet, thereby to estimate the spray efficiency and drift risks. The most famous droplet trajectory models are AGDISP[29] and AgDRIFT[30]. The AgDRIFT was developed for implementation of AGDISP program with the purpose of assessing the off-target pesticide in agricultural application. The development of AgDRIFT was based on the AGDISP. AGDISP Aerial and Ground models can put the spray dynamics into spray fate and drift context for each unique application scenario (e.g. for decision-making) and reasonable worst case scenarios (for risk assessment). The database of these values was supplemented by new formulation, developed nozzles, etc., through experiments. The AGDISP and AgDRIFT are very similar on many aspects, for instance, they have the similar construction, menu items, processing time and interface, except that the AgDRIFT® has three tiers while AGDISP just has one. Indeed, the tier Ⅲ of AgDRIFT is almost equivalent to AGDISP (see Fig. 5), same result could be produced by AGDISP and AgDRIFT from the same parameters input.

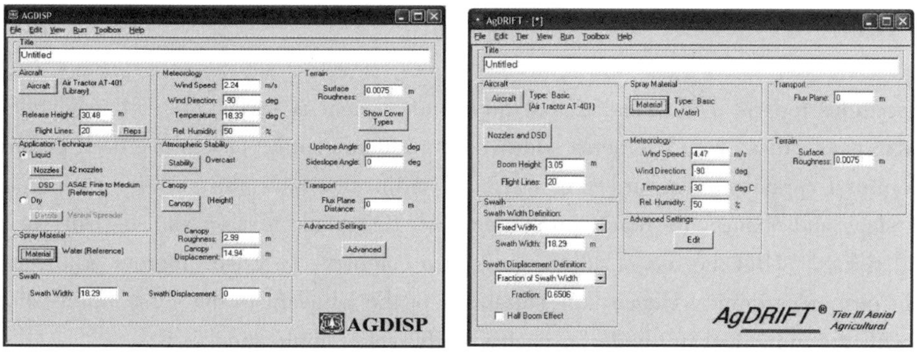

Figure 5 Interfaces of AGDISP and AgDRIFT models

3.2 Atmospheric Dispersion Models

Atmospheric dispersion models were developed to calculate the displace distance and deposition of drops cloud. These models could obtain pesticide concentrations at any geographic position from various atmospheric condition factors including temperature, air relative humidity, wind direction, speed, stability, etc., and estimate spray drift risk. Among these, the most popular model is the "Gaussian plume". Gaussian modelling is a classical approach for atmospheric dispersion modelling, which is suitable for long and medium range drift based on the wind speed and atmospheric stability. [31] The Gaussian model is mainly used for drift prediction in aerial spraying. Other models which were developed by Craig et al. [32, 33] and Raupach et al. [34] used obtained validation results showed good correlation with measurement of downwind deposits for different droplet sizes and wind conditions in aerial spraying.

4 Tank Mix and Other Approaches

The spray quality could be influenced by the physical (liquid density, surface tension and viscosity) and chemical (active ingredients) properties of spray liquid, which depend on tank mixtures (formulation and adjuvants). Moreover, the influence of the physical properties of the liquid on droplet formulation and break-up could not be separated from the mechanical factors of the construction of the spray atomizers. That is why the completely understanding of the influence of spray liquid on droplet size is so difficult. There are also some other approaches, including meteorological parameters (such as temperature, air relative humidity, wind speed and direction), adjustment of spray boom height, vertex migration, wing tip modification devices, and reversing of venturi chamber could help to reduce the spray drift in PAPA.

Not all sprays are equal for a given droplet size classification. Medium sprays with no "fines" can drift less than much coarser sprays with more "fines" and even the coarse droplet size category can still include 1 out of every 10 gallons in "fines". That is why we need to do the math for spraying large areas to find out why eliminating fines is a key DRT.

4.1 Tank Mix

Numerous reports recommended that adjuvants (such as surfactant, oils, polymers and other macromolecules, etc.) to be as additives in pesticide sprays to improve their performance in agricultural spraying in a variety of ways.[35] Adjuvants have the abilities to increase the droplet size, and change the liquid properties such as surface tension and viscosity to reduce the drift risks[36]. But the use of adjuvants is also complex, because different adjuvants fit to different pesticides liquid. Hence, the suitability of the adjuvant to the spray liquid should be tested with the individual pesticide for each specific spray application.

Fig. 6 and 7 showed that emulsion adjuvants tend to enhance break-up of spray sheets, producing narrower droplet size spectra and often fewer fines than solution adjuvants for the same dynamic surface tension.

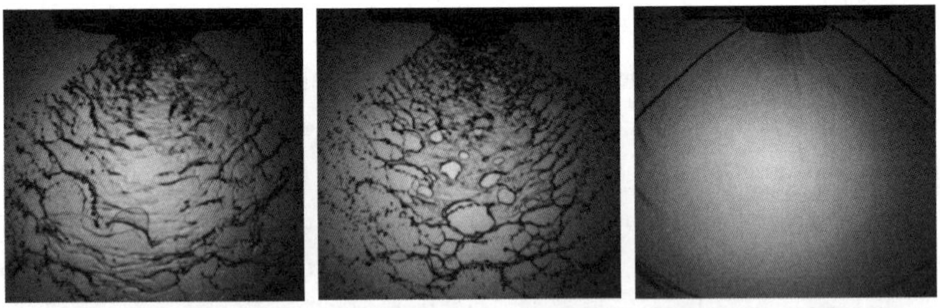

(a) Round up (b) Emulsion (c) Polymer

Figure 6 Different additivities sprays from flat-fan nozzle

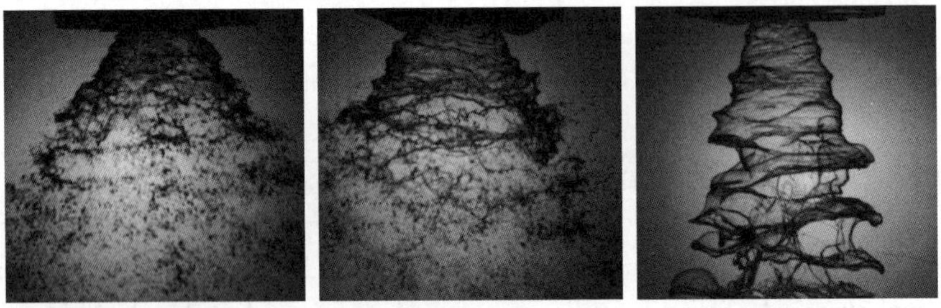

(a) Round up (b) Emulsion (c) Polymer

Figure 7 Different additivities sprays from disc-core nozzle

4.2 Other Approaches

Studies show that optimizing boom placement (lowered boom) after aircraft take-off can avoid high vortex and high air speed areas, and spray drift can be significantly reduced in this way. Rice herbicides are sprayed using Soluble Chemical Water Injection in Rice Technique (SCWIIRT) from two large orifice nozzles trailed below the boom to release the spray less than 0.5m above the target and further away from aircraft vortices (boom length is effectively 35%), drift can be reduced up to 90%. (see Fig. 8)

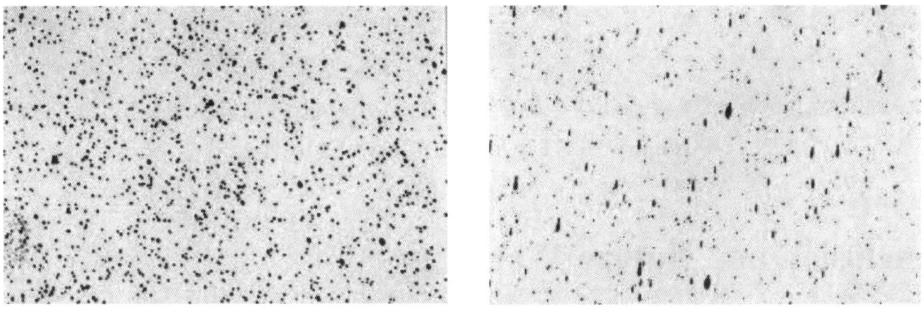

(a) Droplet deposition from lowered boom (b) Droplet deposition from standard boom

Figure 8 Droplet deposition from lowered boom and standard boom

The spray efficiency and drift are related closely to meteorological parameters. Normally, the temperature influences the movements and the relative humidity of the air, and the increase of the temperature is normally consistent with the decrease of relative humidity. And high temperature could lead to the water-based formulations and mixture to be exposed to great evaporation and then result in the formation of small droplets and therefore great drift potential. (see Fig. 9)

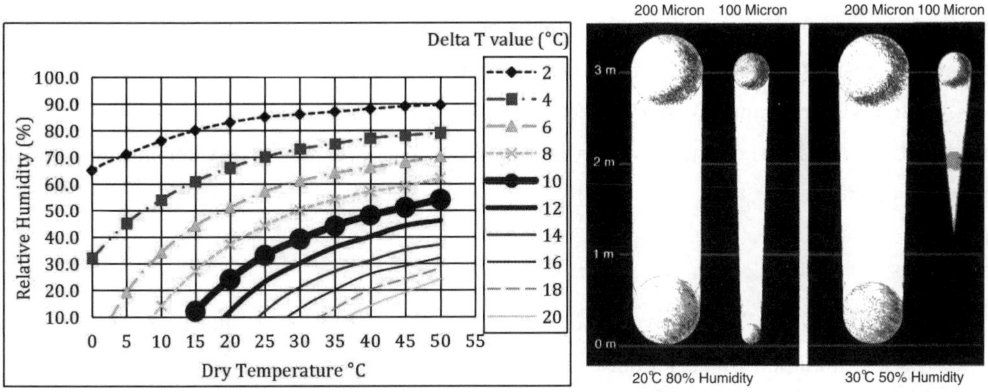

Figure 9 The relationship between temperature and droplet size and relative humidity

When the aircraft lifts, it is an unavoidable consequence that there would be a pair of counter-rotating vortices, which is called wake vortex. These vortexes can cause significant influence on droplet distribution in aerial spray. It makes the vortex (see Fig. 10) mitigation technology to be another main research topic for spray drift reduction.

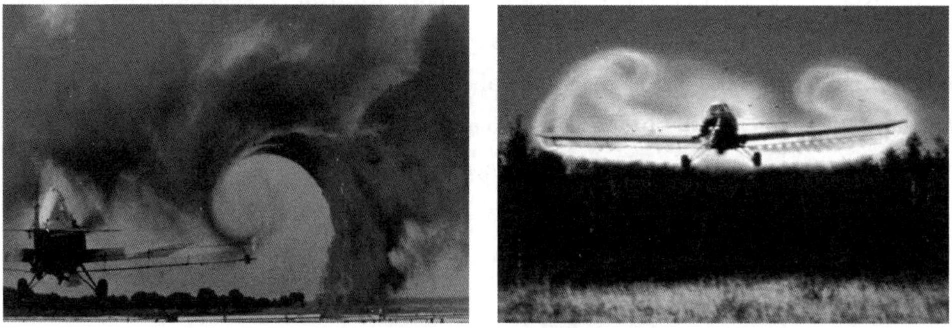

Figure 10 The vortex of aerial spraying

5 Conclusion and Future Work

In agricultural spraying, droplet size is a key factor affecting drift potential and spray efficacy. But the reductions of chemical use and environmental loading in total may be achieved in some cases using finer spray drift management strategies such as physical or vegetation shields, optimized weather and atmospheric stability conditions and other DRTs. Precision aerial application has come a long way through improvements to designs of nozzles, tank mixes, boom placement, real-time control over the spray and possibilities for variable rate and spot/patch spraying.

Considerable work remains to optimize the technologies using sensors and precision aircraft configuration and operation. In the future work for drift mitigation in PAPA, agricultural spraying aircraft need optimization for precision application, especially boom placement and new nozzle designs. Moreover, unmanned aircraft has advantages for precision spraying of spots, patches and bands using detectors/sensors for the target such as weed seekers. Unmanned aircraft for scouting for pests and diseases can provide facility for auditing the pest control efficacy.

References

[1] DORR G J, ANDREW J H, ADKINS S W, et al. A comparison of initial spray characteristics produced by agricultural nozzles [J]. Crop protection, 2013, 53: 109 - 117.

[2] RADCLIFFE J C. Pesticide use in Australia [C]. Parkville, VIC: Australian academy of technological sciences and engineering, 2007.

[3] AL HEIDARY M, DOUZALS J P, SINFORT C, et al. Influence of spray characteristics on potential spray drift of field crop sprayers: a literature review [J]. Crop protection, 2014, 63: 120 - 130.

[4] GIL Y, SINFORT C. Emission of pesticides to the air during sprayer application: a bibliographic review [J]. Atmospheric environment, 2005, 39 (28): 5183 - 5193.

[5] HILZ E, VERMEER A W P. Spray drift review: the extent to which a formulation

can contribute to spray drift reduction [J]. Crop protection, 2013, 44: 75 -83.

[6] VAN DEN BERG F, KUBIAK R, BENJEY W, et al. Fate of pesticides in the atmosphere: implications for environmental risk assessment [J]. Air & soil pollution, 1999, 115 (1 -4): 5 -19.

[7] TONA E, CALCANTE A, OBERTI R. The profitability of precision spraying on specialty crops: a technical-economic analysis of protection equipment at increasing technological levels [J]. Precision agriculture, 2017: 1 -24.

[8] BUENO MARIANA R, CUNHA JOÃO PAULO A R, DE SANTANA DENISE G. Assessment of spray drift from pesticide applications in soybean crops [J]. Biosystems engineering, 2017, 154: 35 -45.

[9] BIRD, S L, PERRY S G, RAY S L, et al. Evaluation of the AgDISP aerial spray algorithms in the AgDRIFT model [J]. Environmental toxicology and chemistry, 2002, 21 (3): 678 -681.

[10] HEWITT A J. Droplet size spectra classification categories in aerial application scenarios [J]. Crop protection, 2008, 27 (9): 1284 -1288.

[11] LIU H, ZHU H, SHEN Y, et al. Development of digital flow control system for multi-channel variable-rate sprayers [J]. Transactions of the ASABE, 2014, 57 (1): 273 -281.

[12] WANG P, LAN Y, LUO X, et al. Integrated sensor system for monitoring rice growth conditions based on unmanned ground vehicle system [J]. International journal of agricultural and biological engineering, 2014, 2 (7): 75 -81.

[13] LORENS J, GIL E, LLOP J, et al. Ultrasonic and LIDAR Sensors for electronic canopy characterization in vineyards: advances to improve pesticide application methods [J]. Sensors, 2011, 11 (2): 2177 -2194.

[14] BERK P, HOCEVAR M, STAJNKO D, et al. Development of alternative plant protection product application techniques in orchards, based on measurement sensing systems: a review [J]. Computers and electronics in agriculture, 2016, 124: 273 -288.

[15] LAN Y, HUANG Y, MARTIN D E, et al. Development of an airborne remote sensing system for crop pest management: system integration and verification [J]. ASABE, 2009, 25 (4): 607 -615.

[16] VASHAHI F, RA S, CHOI Y, et al. A preliminary investigation of the design parameters of an air induction nozzle [J]. Journal of mechanical science and technology, 2017, 31 (7): 3297 -3303.

[17] THOMSON S J. Evaluation of a solid stream radial nozzle on fixed-wing aircraft, for penetration of spray within a soybean canopy [J]. Journal of plant protection research, 2014, 54 (1): 96 -101.

[18] MEISCH M V, DAME D A, BROWN J R. Aerial ultra-low-volume assessment of ANVIL 10 + 10 against Anopheles quadrimaculatus [J]. Journal of the American mosquito

control association, 2005, 21 (3): 301 -304.

[19] HEWITT A J, ROBINSON A G, SANDERSON R, et al. Comparison of the droplet size spectra produced by rotary atomizers and hydraulic nozzles under simulated aerial application conditions [J]. Journal of environmental science & health part B, 1994, 29 (4): 647 -660.

[20] VALLET A, TINET C. Characteristics of droplets from single and twin jet air induction nozzles: a preliminary investigation [J]. Crop protection, 2013, 48: 63 -68.

[21] VASHAHI F, RA S, CHOI Y, et al. Influence of design parameters on the air/liquid ratio of an air induction nozzle [J]. Journal of mechanics, 2017: 1 -11.

[22] MINOGUE P. Advances in aerial application technology [C/OL]. Redding, CA: Proc. 25th Ann. For. Veg. Mgt. Conf., 2004. (http://www.fvmc.org/PDF/2004/25%20FVMC%20Proceedings%202004.)

[23] THOMPSON J C, ROTHSTEIN J P. The atomization of viscoelastic fluids in flat-fan and hollow-cone spray nozzles [J]. Journal of non-newtonian fluid mechanics, 2007, 147 (1): 11 -22.

[24] WEI Z, HOU Y R, LIU X, et al. Wind tunnel experimental study on droplet drift reduction by a conical electrostatic nozzle for pesticide spraying [J]. International journal of agricultural & biological engineering, 2017, 10 (3): 87 -94.

[25] PATEL M K, PRAVEEN B, SAHOO H K, et al. An advance air-induced air-assisted electrostatic nozzle with enhanced performance [J]. Computers & electronics in agriculture, 2017, 135 (C): 280 -288.

[26] PATEL M K. Technological improvements in electrostatic spraying and its impact to agriculture during the last decade and future research perspectives: a review [J]. Engineering in agriculture environment & food, 2016, 9 (1): 92 -100.

[27] TESKE M E, BIRD S L, ESTERLY D M, et al. AgDRIFT: a model for estimating near-field spray drift from aerial applications [J]. Enviromental toxicdogy & chemistry, 2002, 21 (3): 659 -671.

[28] BILANIN A J, TESKE M E, BARRY J W, et al. AGDISP: the aircraft spray dispersion model[J]. Code development and experimental validation. 1989, 32(1):0327 -0334.

[29] TESKE M E, BIRD S L, ESTERLY D M, et al. AgDRIFT: a model for estimating near-field spray drift from aerial applications [J]. Environmental toxicology & chemistry, 2002, 21 (3): 659.

[30] BACHE D H, SAYER W J D. Transport of aerial spray: a model of aerial dispersion [J]. Agriculture meteorol, 1975, 15: 257 -271.

[31] CRAIG I, WOODS N, DORR G. A simple guide to predict aircraft spray drift [J]. Crop protection, 1998, 17 (6): 475 -482.

[32] CRAIG I P. The GDS model: a rapid computational technique for the calculation

of aircraft spray drift buffer distances [J]. Computers and electronics in agriculture, 2004, 43: 235 – 250.

[33] RAUPACH M R, BRIGGS P R, AHMAD N, et al. Endosulfan transport Ⅱ: modelling airborne dispersal and deposition by spray and vapour [J]. Journal of environmental quality, 2001, 30: 729 – 740.

[34] HEWITT A J, MILLER P C H, BAGLEY W E. Evaluation of the effects of adjuvants on agricultural spray characteristics [C]. Pasadena, CA: Proceedings 8th international conference on liquid atomization and spray systems, 2000.

[35] HEWITT A J, STERN A J, BAGLEY W E, et al. The formation of a new ASTM E35. 22 task group to address drift management adjuvants [C]. West Conshohocken, PA: 19th symposium on pesticide formulations and application systems: global pest control formulations for the next millennium, 1999. (American society for testing and materials.)

[36] PISC. Spray drift management. Principles, strategies and supporting information [R]. Primary industries standing committee. PISC (SCARM) Report, 2002: 82.

作者简介

安德鲁·约翰·休伊特教授（Professor Andrew John Hewitt），澳大利亚昆士兰大学科学学院农业与食品安全系高级研究员，农药施用与安全中心主任，美国内布拉斯加大学兼职教授。主要研究和教学领域包括：农药在农业和工业生产中的使用、漂移、雾化、扩散、沉积，对环境的影响，喷施优化，对农药施用效能评估以及森林疾病防控与治理等。目前主持的主要研究项目包括：由谷物研究与发展公司资助的粮食作物病虫害防治项目，研究经费为 210 万澳元；由澳大利亚葡萄和葡萄酒管理局资助的葡萄病虫害防治项目，研究经费为 200 万澳元；另外还有来自澳大利亚其他公司、组织及政府的资助，如澳大利亚园艺创新与发展公司、棉花研究与发展公司等。2013 年，作为美国国土安全协调研究与合作科学部门的代表，以鉴定专家的身份，参加了国际法庭关于哥伦比亚与厄瓜多尔农药漂移案的审判。作为国际农业标准组织主要成员，致力于国际农业相关标准的制定，如施药过程中的施药效能和施药漂移鉴定标准等。此外，他还是三家国际期刊的编辑委员，*TRANS of the ASABE* 和 *Pest Management Science* 的副主编，多个国际技术委员会的主席和多个国际学术会议的组织者。多年来，他一直致力于农药喷施的风洞实验室以及田间试验研究，建立雾化漂移预测模型，并开展澳大利亚政府机构及工业组织施药技能以及雾化漂移预测模型使用培训。

廖娟（Juan Liao），博士，华南农业大学助理研究员，湖南省益阳市人。从事航空植保施药及基于病虫害监测的作物精准施药等研究工作。采用多旋翼无人直升机喷施棉花脱叶剂，优选了作业参数，相比传统施药方式，提高了作业质量和作业效率，减小了对棉花植株的损伤以及对环境的污染；基于 LiDAR 监测作物冠层和 PWM 控制喷嘴的智能变量施药系统，对葡萄病虫害进行防治，相比传统均匀施药方式，提高了施药质量，减少了施药量和环境污染；基于多传感器融合和深度学习技术，对水稻褐飞虱进行监测，提高了对褐飞虱监测的精度。

典型农用无人机研发与应用

何　勇　岑海燕①

1　引言

农用无人机是应用于农业中的固定翼、单旋翼、多旋翼和热动力气球等小型低空无人航空作业机械的统称，是一种有动力、可控制、能携带多种任务设备、执行多种任务，并能重复使用的航空器。近年来，无人机在农业中的应用越来越广泛，主要集中在农田信息遥感监测[1]和植保喷药[2]两个领域。

1987 年，世界上第一台农用无人机出现。日本 Yamaha 公司受日本农林水产省委托，生产出 20 千克级喷药无人机“R-50”。[3]我国系统研究微小型无人机航空施药喷雾技术开始于 2008 年，由国家资助，开始单旋翼无人机低空低量施药技术的研究。无人机植保作业相对于传统的人工喷药作业和机械装备喷药具有较多优点：作业高度低，飘移少，可空中悬停，无需专用起降机场，旋翼产生的向下气流有助于增加雾流对作物的穿透性，防治效果高，远距离遥控操作，避免了喷洒作业人员暴露于农药的危险，提高了喷洒作业安全性等。无人直升机喷洒技术采用喷雾喷洒方式至少可以节约 50% 的农药使用量，节约 90% 的用水量，这将在很大程度上降低资源成本。21 世纪以来，随着计算机技术、导航技术、控制技术和通信技术的发展，以及各种体积小、重量轻、检测精度高的便携式传感器的出现，无人机低空遥感技术在农业中的应用逐年增多。[4]农用无人机低空遥感相比于传统的卫星和航空遥感，在时效性、准确度、可操作性、成本以及对复杂农田环境的适应性等方面有显著的优势，已成为现代农业信息技术的研究热点和未来的主要航空遥感技术之一。农用无人机低空遥感技术完善了遥感技术的时空分辨率，为多维度信息的准确获取、实现农业的精准化管理和决策提供技术支持。

2　农用无人机专用机型

农用无人机有多种分类方法：按动力来源可分为电动和油动，按机型结构可分为固定翼、直升机、多旋翼，按起飞方式可分为助跑起飞、垂直起飞垂直降落，按功能不同可分为农事操作和农田信息采集，等等。[5]以下主要介绍油动直升机和多旋翼无人机两

①　何勇、岑海燕，浙江大学生物系统工程与食品科学学院。

种农用无人机机型。

2.1 油动直升机

油动直升机具有荷载大、操作灵活、使用时间长、作业效果好等优势。其通过主桨切割空气产生推力，尾桨保证平衡，无需跑道助跑，可垂直起降和稳定悬停，可靠性强，操作灵活度高，可快速上升下降，搭载载重高。目前，国内最大起飞重量和最大载重的无人机就是采用直升机结构。油动直升机一般由机体、发动机、传动、旋翼和尾杆等部件约200多个机械零件组成，并搭载电子系统，可实现飞行过程中的稳定控制和自主飞行。油动直升机的技术发展方向有：①发动机。目前，国内多采用二冲程化油器式发动机。温度、载荷和气压变化都会直接影响发动机工作。国内在用的大多是航模发动机而非航空发动机，动力输出不稳定，不符合通用航空的发动机标准，也难以取得适航证，需研发适用于农用无人机的专用航空发动机。②旋翼。目前，国内多采用常规航模直升机旋翼，很少针对农用无人机的进行设计，在大载重飞行时，为保证升力足够，需要增加旋翼面积；如果只延长旋翼长度，在相同转速下翼尖线会因速度接近或超过音速而失效。这需要根据飞机动力配置情况，旋翼翼尖和翼根采用不等弦长设计。如Yamaha公司的Rmax，根据自身的动力输出，专门设计旋翼翼型，依据仿真受力效果，设计独有翼型，这既保证了强度，又提高了升力。③农药喷头。目前，国内采用传统喷头，这不能满足航空喷药需求。我们需要改变弥雾方式，设计气流进行喷雾，并根据不同作物和药剂开发相应的专用航空喷头。④安全纪录仪。黑匣子数据可以用于事故分析、定性事故类型。目前仅有部分飞机有飞行姿态记录，传感器数据量少，例如，发动机温度、空速等传感器没有被安装，这不便于进行事故分析。我们需要开发飞行数据的记录仪，这便于事故分析和认定。⑤云台。目前，国内多采用航拍云台，这较难适应精细农业信息采集的要求。我们需要开发农用信息获取专用云台，如专业的光电吊舱等高精度云台。

2.2 多旋翼无人机

多旋翼无人机以三个或者偶数个对称非共轴螺旋桨产生推力上升，以各个螺旋桨转速改变带来的飞行平面倾斜实现前进、后退、左右运动，以螺旋桨转速次序变化实现自转，垂直起飞降落，场地限制小，可空中稳定悬停。多旋翼飞行器出现后，以优越的飞行稳定性、简单的动力学结构和低廉的价格迅速获得广泛的关注和使用。多旋翼无人机采用锂聚合物电池供电，自动化程度高，飞行平稳，操作技术要求低。同时，由于结构所限，载重量一般不高（10千克左右），续航时间较短（15～20分钟）。多旋翼无人机的技术发展方向有：①一体机。目前，多数多旋翼无人机采用组装、拼凑的方式，缺乏科学的系统设计，单一追求载重，盲目增加旋翼数目来提高承载能力。通过设计一体机将多旋翼与植保喷药深度整合，提高整机性能，采用大扭矩电机，携带喷杆，用户只需要更换电池，注入药液即可。②机身强度。目前，多旋翼多采用碳纤维作为骨架，其模量低，强度小，在频繁起飞的过程中容易开裂，影响骨架强度和准直度；多采用碳管直接切割制作，缺乏针对受力分布的合理设计；机臂在飞行时，受力并不是均匀分布，根部存在应力集中现象，机臂需要科学设计。③药箱设计。药箱设计不合理，在喷药过程中，无人机进行转向，加、减速等动作时，会出现滞机现象，稳定性差。我们需要合理设计药箱布局，合理分散重量。④电池。开发专用电池，提高电压、降低电流，这样可以减少能量损耗和机身损伤，油电混合动力也是一种发展趋势。⑤电机。电机采用开放设计，会混入杂草残骸和昆虫尸体等杂质，增大电机摩擦，缩短电机寿命。我们需要开发专用的大扭矩节能型变频电机。

3 飞行控制系统

目前，大多数无人机采用商品飞控系统，国外如加拿大 Micropilot、德国 Helicomand 等，国内如大疆飞控、零度飞控等，但匹配性差，不能很好地满足农业作业需求，会造成飞行过程中的不稳定性和存在安全隐患。在农业中应用的无人机，其作业环境与其他民用领域的无人机有较大的区别。比如，在喷药作业中，为了实现药剂的精准变量喷洒，农用植保无人机通常需要以较慢的巡航速度在较低的飞行高度下进行喷洒作业。农业专用飞控系统就需要针对低空风切变等作业干扰因素进行及时的处理与修正。而在无人机低空遥感作业中，由于需要对遥感图像进行几何校正等处理，所以飞控系统需要提供精度更高的飞机位置与姿态数据对遥感数据进行修正。总体而言，农用无人机飞控系统与通用飞控系统的最大区别在于针对农业特殊的作业环境，对飞控系统的可靠性、安全性提出了更高的要求。应当开发专用于农业作业的无人机飞控系统，优化控制算法中的各项参数，加入自适应算法，提高无人机在田间恶劣环境下作业的适应性和稳定性。目前，农用无人机需解决的飞行控制关键技术有以下几项。

3.1 作业路径规划与精确导航技术

导航信息采用差分全球定位系统和北斗卫星导航系统并结合惯性导航系统做捷联解算，提高在动态环境下的高精度位置信息获取效率。设定作业田块的边界、喷洒速度和喷幅，即可自动规划航线，作业时按照已规划的航线实行全自主飞行作业以减少重复喷洒带来的药害和浪费，对飞行速度的管控可以保证喷施的均匀。能根据不同的地形、地貌规划最佳的作业路径，实现自主飞行，最大限度地减少目前目视操作带来的重复作业和遗漏作业。

3.2 障碍物自动感知与避障技术

目前，避障技术主要基于超声波测距、激光雷达、红外测距、双目立体视觉及微波雷达等手段，它们都存在着各自的弊端。我们需在飞行器上研制针对树干、电线杆等农村常见障碍物的感知与避让装置，实现自适应巡航速度控制，在应对突然出现在飞行路径上的物体时可立即阻断飞行或做出合理规避，保障飞机和障碍物的安全。

3.3 一键起飞与自主降落技术

通过分析农用小型无人机对地面控制站的需求，一键起飞和自主降落技术在地面控制站的结构和功能实现方法，分析开发地面控制站涉及的关键技术和解决方案。地面控制站具有实时性强、稳定性好、人机界面友好和可扩展能力强等优点；并能实现断电续航及智能化操控，因续航时间问题需更换电池时上电重启后执行航点续飞功能。地面控制站支持任意多边形飞行区域的覆盖式作业航线规划，操作简单，选中区域且进行简单配置，如航点间距或作业半径等即可生成航线。

3.4 仿地形飞行技术

为了适应山地、丘陵等不同地貌以及不同高度的作物作业需求，达到最佳的作业效果，农业无人机需要仿地形飞行。在飞行过程中实时感知对地绝对高度，使用外部多种高度传感器融合技术结合内部惯性导航传感器，分析地面与飞机的实际高度，减少植物冠层对高度的影响，实现全自动仿地形飞行。

4 浙江大学微小型无人机遥感技术与平台

4.1 自主研发无人机技术

4.1.1 多旋翼无人机

浙江大学自主研制的八旋翼无人机，机身直径 1.1 米，高 0.35 米，使用最先进的锂离子聚合物电池，最大续航时间为 25 分钟，最快飞行速度达到 75 千米/小时，最高可飞 500 米，可挂载 8 千克的设备或药物，使用卡尔曼滤波和 I2C 交流电机驱动器，攻克了飞行器在不同天气条件、不同作业载荷下稳定飞行的技术难关。特别是喷洒作业时，所搭载的药量不断减少，但飞行器始终保持重心位置不变，此技术获得了 4 项专利。该无人机具有以下优点：第一是有很好的安全性，在丢失遥控信号及电池电量即将用尽的情况下自主返回降落，在一只机臂失效的情况下，仍能安全飞行，最大限度减少意外情况对无人机造成的损伤，减少经济损失。第二是质轻便携，全身碳纤结构，机身仅重 3.5 千克，可折叠，便于携带和运输。第三是用途广泛，可搭载 RGB 相机和光谱相机实现农田信息采集，如图 1（a）所示；也可搭载喷药装备在农田喷药，喷药效率大大高于人工喷药，如图 1（b）所示。

（a）低空遥感作业　　（b）植保喷药作业

图 1　浙江大学自行研发的农用八旋翼无人机

4.1.2　油动直升机

图 2 是浙江大学自行研发的农用油动直升机。此油动直升机装备了 170 毫升水冷汽油发动机，机身长 2.75 米，旋翼直径 2.3 米，最大起飞全重 35 千克。全机采用进口航空铝材，整体 CNC 加工成型。该研发团队提出了工业级无人直升机的制造方法，攻克了传统无人机大扭力变速箱寿命短、空中故障率高、发动机易过热等技术难点。采用多片级湿式离合器和液循环冷却变速箱，提高了变速箱的可靠性，同时增加了变速箱的寿命。在发动机方面，采用智能热量管理系统，使发动机可以在稳定的环境温度下工作。在控制方面，飞机自动化程度高，可以在独立自主起降、智能导航飞行与计算机辅助手动操作之间无缝切换。

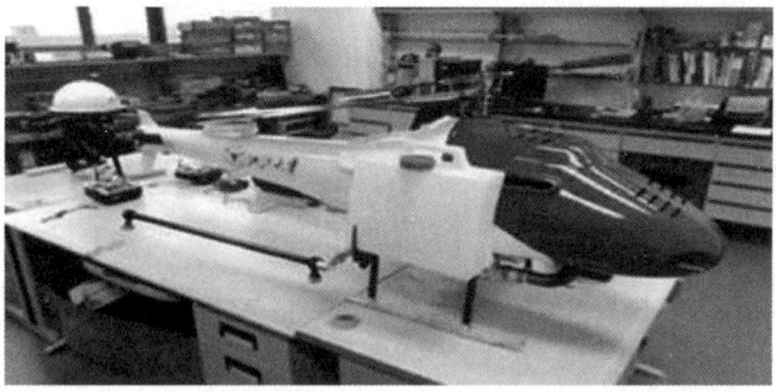

图 2　浙江大学自行研发的农用油动直升机

4.1.3　五自由度全飞行姿态仿真模拟平台

图 3 为浙江大学研制的五自由度全飞行姿态仿真模拟平台。该无人机模拟系统模拟无人机在农田中行走时位置和姿态的变换，设计包括水平向自由度、垂向自由度、俯仰、横滚、偏航 5 个自由度。[6]该系统设计分为 3 个层次，首先是底层部件包括水平和垂直伺服电机驱动器、接近开关、远端执行器。其中，远端执行器为次级控制器，向下获取传感器信息并输出控制指令，以此实现不同控制小区的通信。通信部分存在于不同层之间以及远端控制器与内部机械部分、控制器部分和上位机软件，三部分分别进行开发。如图 3（b）所示，该系统包括水平和垂直两个方向的导轨，实现对水平位移速度和垂直喷洒高度的控制，采用两个伺服电机实现对水平移动速度和垂直移动速度的精确控制。

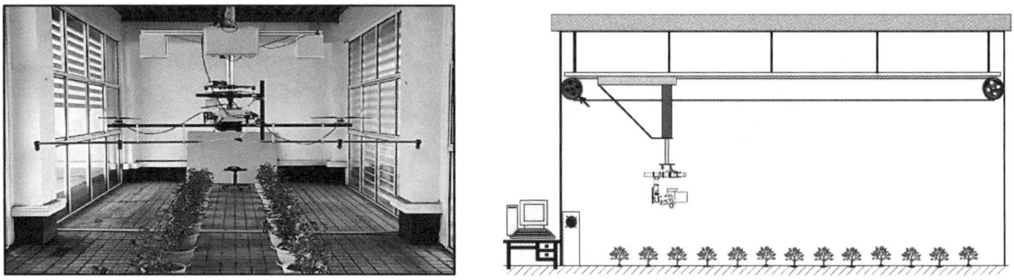

（a）场景　　（b）结构

图3　五自由度全飞行姿态仿真模拟平台

4.2　自主研发飞控系统

图4所示为浙江大学自行研发的农用无人机飞控系统。首先，图4（a）飞控板采用了DSP与ARM的双MCU（微控制单元）结构，通过中央MCU的PID闭环控制实现前进、后退、升降等动作。该设计保证了飞控系统运算的精确性和实时性，能够控制飞机对飞机姿态与外界飞行环境的变化做出及时和有效的响应，从而维持飞机在低空复杂环境中飞行的稳定，提高了安全性。其次，在飞控系统中安装了高可靠性的MEMS三轴陀螺仪、加速度计、三轴磁感应器以及气压传感器等装置。结合外部高精度GPS模块，其可以为无人机农业作业提供准确的经纬度、飞行高度和速度等导航姿态数据。再次，由于农用无人机经常在作业情况较为恶劣的地区进行飞行，故飞控系统的电路接插件设计摒弃了常用飞控中的杜邦线与排针的组合方式而采用了防水与防尘性能更高的航空插头。另外，因为与通用无人机相比，农用无人机需要频率更高的维护和保养，所以，在农业专用飞控系统中设计可靠的数据存储装置，用以在每次飞行中对飞机各种姿态信息、传感器信息、电气系统参数、位置信息和任务载荷信息进行实时记录，方便对无人机飞行参数与状态进行实时记录与监控，为农用无人机的维护保养提供可靠的信息与依据。该飞控系统可实现高精度定位、智能避障技术、仿地飞行、断电续航、智能化操控、任意多边形航线规划［见图4（b）］和故障自检等功能。

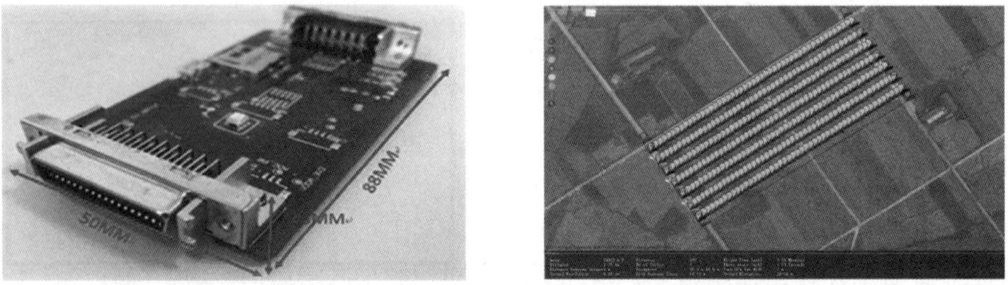

（a）无人机飞控板　　（b）自动航线生成

图4　浙江大学自行研发的农用无人机飞控系统

4.3 基于无人机遥感的农田信息采集

4.3.1 基于无人机三维空间运动模拟平台的作物信息采集

五自由度全飞行姿态仿真模拟平台可以模拟无人机的空中航姿变化，搭载多种传感器，实现了可控环境下无人机动态信息采集与处理。在导轨平台上模拟无人机在空中往复运动采集作物的 RGB 图像，将所得图像采用运动恢复结构的算法进行三维重构，依据网格化重建后的曲面进行面积计算，得到每个细小网格的面积，之后进行累加，可以得到作物的体叶面积[7]，如图 5（a）所示。在模拟平台上搭载 PMD 深度相机和 TetracamADC 多光谱相机[8]。多光谱相机应用于区分地物，PMD 深度相机获得植物高程三维信息，以此获得植物营养三维空间分布，采用 Powell、粒子群优化算法和遗传算法的图像匹配融合，可以将两幅非同源图像依据最大互信息进行融合，以此获得植物营养三维空间分布，如图 5（b）所示。在模拟平台上搭载 Tetracam ADC 多光谱相机，利用 soil and plant analyzer development（SPAD）值作为油菜含氮量的评判标准，基于多光谱遥感图像获得植被指数和图像纹理特征，建立油菜冠层 SPAD 值的预测模型，并探究油菜冠层多光谱图像获取时间、相机高度、运动速度对预测模型的影响，然后利用基于植被指数建立的预测模型对 SPAD 值可进行可视化反演[9]，如图 5（c）所示。

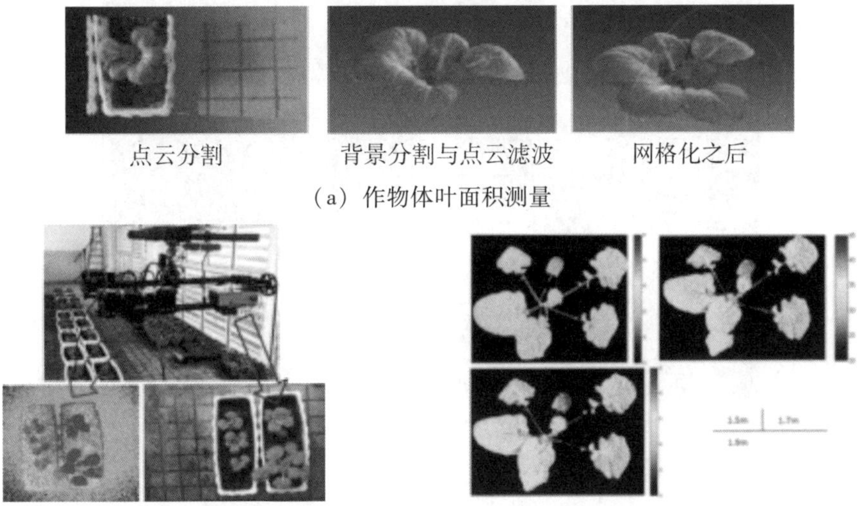

点云分割　　背景分割与点云滤波　　网格化之后

（a）作物体叶面积测量

（b）基于多传感器融合的作物营养三维分布　　（c）油菜冠层可视化分析

图 5　基于无人机三维空间运动模拟平台的作物信息采集

4.3.2 基于光谱成像技术的无人机遥感平台研发和应用

基于光谱成像技术的无人机低空遥感平台，可实现图像的零延时高清回传以及遥感影像的辐射校正和几何校正，准确、有效地建立作物养分、长势遥感反演模型。无人机低空遥感平台由三轴云台、多光谱相机系统、RGB 相机及图传等组成。多光谱相机系统为 CMOS 多光谱图像传感器全幅成像，有 2048 × 1080（两百多万像素）的分辨率和 42 fps 的输出帧率，25 个波段（600 ～ 1000 nm）。通过计算机控制多光谱相机抓拍和存储图像，还可实现对多光谱相机增益及曝光时间的调节。RGB 相机为索尼 Sony NEX-7 数码相机。数字图传的空中端重量仅 130 克，发射频率 5.8 GHz，有效传输距离 1 千米，可以实现图像传输的零延时。云台为三轴无刷云台，可以同时触发 RGB 相机和多光谱相机拍照，还配备了辐射定标系统和地面控制点。

4.3.3 基于无人机低空遥感的油菜养分信息获取

试验地点为浙江大学紫金港西区，试验地块土质为中壤土，肥力上等，采用常规大田管理方法管理油菜植株，油菜品种为浙大 630。为充分研究氮素对油菜生长状况的影响，我们只对氮肥进行梯度施肥的试验。不施、少量、适量和过量的比例遵照 0∶0.5∶1∶1.5；氮肥试验的所有小区均施用适量的磷肥和钾肥，保障油菜植株对磷肥和钾肥的生长需要。规划好无人机航线后，利用低空遥感平台按航点获取油菜地的 RGB 图像和多光谱图像，对图像进行辐射定标和几何校正，再按图像序列对图像进行自动拼接，获取整块油菜田地的正射图。在正射图中以小区为单位，从光谱变量、植被指数以及图像特征的角度建立大田油菜冠层的养分预测模型，选出最优模型后进行可视化反演（见图 6）。由此得到的大田油菜的养分分布情况对后续的施肥作业起到了关键的指导作用。

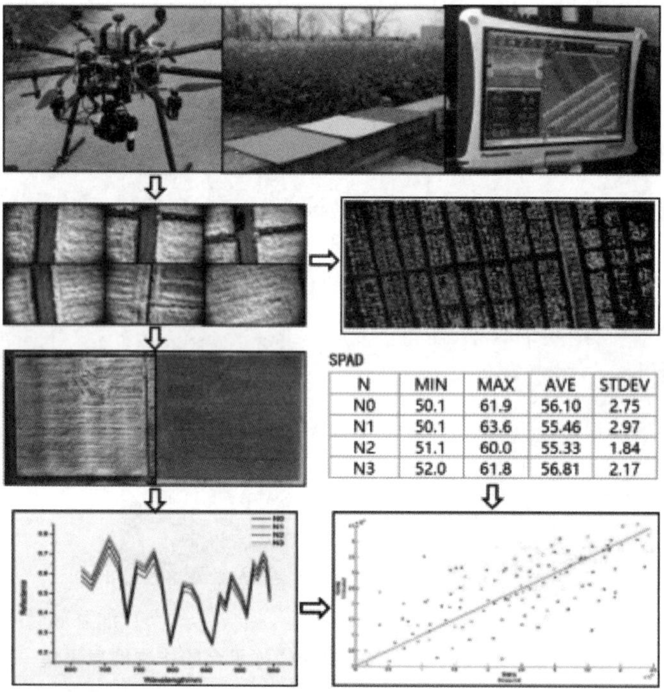

图 6　基于光谱成像技术的无人机遥感平台及其在油菜养分信息获取中的应用

参考文献

[1] 杨贵军，李长春，于海洋，等. 农用无人机多传感器遥感辅助小麦育种信息获取 [J]. 农业工程学报，2015，31 (21)：184－190.

[2] 杨益军. 中国植保无人机市场现状和前景分析 [J]. 农药市场信息，2015

(13)：6－9.

［3］何勇，张艳超. 农用无人机现状与发展趋势［J］. 现代农机，2014（1）：1－5.

［4］范承啸，韩俊，熊志军，等. 无人机遥感技术现状与应用［J］. 测绘科学，2009，34（5）：214－215.

［5］杨陆强，果霖，朱加繁，等. 我国农用无人机发展概况与展望［J］. 农机化研究，2017（8）：6－11.

［6］ZHANG Y，XIAO Y，ZHUANG Z，et al. Development of a near ground remote sensing system［J］. Sensors，2016，16（5）：648.

［7］张艳超，庄载椿，肖宇钊，等. 基于运动恢复结构算法的油菜 NDVI 三维分布［J］. 农业工程学报，2015，31（17）：207－214.

［8］张艳超，肖宇钊，庄载椿，等. 基于小波分解的油菜多光谱图像与深度图像数据融合方法［J］. 农业工程学报，2016，32（16）：143－150.

［9］肖宇钊. 基于低空光谱成像遥感技术的油菜冠层 SPAD 检测研究［D］. 杭州：浙江大学，2016.

作者简介

何勇，浙江大学求是特聘教授，现任浙江大学农业信息技术研究所所长、浙江大学数字农业与农村信息化研究中心常务副主任、农业农村部光谱学重点实验室主任，国家重点学科—农业机械化工程学科学术带头人之一、“十二五”国家 863 现代农业领域“数字农业技术与装备”主题专家、863 项目首席专家、国家教学名师、百千万国家级人才、国家农村信息化示范省国家级指导专家。曾获国家级教学团队、浙江省首届十大师德标兵、第四届浙江省十大杰出青年、浙江省优秀杰出青年基金等荣誉，荣获浙江大学永平教学贡献奖。主要从事数字农业、农业物联网、农村信息化、农用航空和智能农业装备等方面的科研和教学工作。曾先后在日本东京大学、东京农工大学、美国伊利诺斯大学访学和担任客座教授，入选汤森路透 2016 年全球高被引科学家（Highly-Cited Researchers 2016）。

Agricultural Aviation: Looking Back, Status, Prospects

Marc Vanacht[1]

In 1995 the Fertilizer Institute, the research arm of the international fertilizer industry at that time published a drawing to explain the concept and expected developments of precision agriculture (PA). It showed GPS navigation, handheld data loggers, yield monitors, soil sampling, maps with different data layers for different aspects of soil quality. However, it did not show satellites for crop observations, planes or helicopters for crop observations and treatments. Many aspects of PA conceived in 1995 have been adopted in the daily practice on large farms in the US, Brazil, NE China, and Australia.

Recent developments in agricultural aviation have opened new possibilities: crop observations and crop treatments by cheap and simple aerial platforms to serve the needs of small to medium farms in densely populated areas, and in crops with special constraints like rice.

What can we expect from agricultural aviation in the years to come?

Crop observations with satellite platforms will evolve. Micro-satellites may improve re-visit frequency but pixel detail remains an issue: 5m pixel is acceptable for big acre, low value field crops, whereas <0.5m pixel is useful for higher value but small acre crops. So there seems to be a disconnect between value and satellite sensors capabilities. Low value crops cannot afford the cost of the satellite data, and high value crops do get the pixel size detail they need to be relevant. The only solution may be a "national" interest investment in satellites, made available to the public like military GPS signals ended up being made available to the wider public. Will countries like China or Brazil make an investment of that kind?

Recently, UAVs also called drones have moved from the realm of "entertainment & photography" into commercial uses. This largely happened because from 2009 to 2016 the cost of a kWh of battery power dropped about 75%. (Sources: Exane BNP Paribas, UBS, *The Economist*). Small drones benefited from this. Miniaturization and automation enabled increased payloads and broader functionality, not just as observation platforms, but also as application platforms for smaller fields. Japan had been using remote control helicopters for treatment in rice since the early 2000s, but this new trend of cheap electric platforms fundamentally changed the

① Marc Vanacht, AG Business Consultants, St Louis, MO, USA.

cost of aerial observations and applications.

The next 10 years, cost reduction in battery power is expected to continue incrementally benefiting further adoption of drones. Today companies in Japan, China and South Korea dominate production of batteries for mobility. Will this change when the Solar City mega plant in Arizona comes on stream? Also, scientific and technical breakthroughs may happen, and new scenarios of agricultural applications of aerial platforms may become possible. The issue is not if, but when?

What is the status of adoption of drones? Figures for the USA were published early April 2017. According to Michael Huerta [Chief of the Federal Aviation Authority (FAA), quoted by Eric Sfiligoj of *Precision Ag Magazine*] between December 2015 (when regulation of drones were initiated) and March 2017, 770,000 drones were registered, 100,000 of which were registered in the first 3 months of 2017. By 2020, FAA expects a total of 3,000,000 registrations. Meanwhile, to date (April 2017), 37,000 commercial "remote pilot" certificates have been issued. These certificates allow the holders to perform commercial activities (like applications in agriculture). The figures for agriculture have not been published or are not known. In China figures were published about 3,000 drones used in agriculture in 2015, and possibly 10,000 units by early 2017. Japan uses around 2,000 drones in agriculture, a figure that has been stable since many years.

With the high level of adoption (at least outside agriculture) some writers wonder whether drones are tools or toys? In a March 15, 2017 article in *Precision Ag Magazine* Ben Johnson notes that the number of companies selling drones services in agriculture dropped from dozens to very few between 2015 and 2017. He also notes weather, technical and business process constraints to transform NDVI images into actionable conclusions. He recognizes however that RGB pictures taken with drones are very useful and allow farmers to track the growth of their crops. In an April 24, 2017 article, Louisa Burwood-Taylor of *Ag Funder News* wonders why investments in drones services for agriculture dropped 64% from 2015 to 2016. She concludes that drones are simply not ready for the "big farming" in the USA for several reasons: battery life, autonomy, payload and image processing.

All that does not mean drones may be ready for other agricultural situations like for example in China, or as proven by nearly 20 years of use, for treatments of rice in Japan. Any situation of small to medium farmers, with difficult applications conditions (hilly geography, wet situations like rice, dense population centers surrounding fields, etc.) will be ready to adapt drones before situations of large fields, flat land, low population densities like in the US Mid West, many parts of Brazil Central Eurasia, Australia, etc.

One of the main uses of drones is for crop observations with the use of sensors. NDVI/ReNDVI (Normalized Differential Vegetation Index-Red Edge, NDVI) multispectral sensors are very much the industry standard. "On the ground" active N sensors are used in proximal applications. In remote applications, distance make active sensors a difficult/impossible target. That means most data from remote sensing (drones, planes, satellites) will remain relative data. Actionable recommendations can only come from additional ground observations. This requires more complicated "business processes" than just flying a drone to take pictures.

We can expect breakthroughs in sensor technologies that would give "absolute measurements" leading to actionable prescriptions. Single chip NDVI, multispectral, hyper spectral sensors will improve the situation. These technologies exist but costs need to come down so farmers can afford their implementation. Also, after capturing the initial "picture", the downstream processing to useful data needs further automation. Taking images (using any platform) and developing prescriptions today is a typical service business, more people intensive than capital intensive. The industry needs to get to <24 hours turn around time to be useful. This means image capture, ground truth, data interpretation, treatment prescription and commands for application machines all need to be done within that timeframe. We are not yet there today, but it should happen in the near to medium future. Medium term "deep learning" artificial intelligence may be promising as help for analysis and recommendations.

In the USA, B. Erickson and D. A. Widmar put forward in the 2015 *Precision Agricultural Services Dealership Survey* published by Purdue University in August 2015 that 48% of agricultural retailers offered satellite image and 19% drones image services. According to the retailers these figures should grow to 55% and 38 % respectively by 2018. We are all curious to see how the figures from the upcoming 2017 survey (may be available August 2017) will end up to be.

The same publication (Erickson and Widmar) shows precision agriculture technology trends in applications. GPS guidance (autosteer) is now used by 83% of retailers, GPS enabled sprayer, boom section and nozzle control by 74% of retailers. About half (51%) use satellite and aerial images for management purposes, and 16% work with drones. Solenoid valves, single nozzle control and map or sensor based variable rate applications are now common by retailers, contractors and by large farmers. As advanced equipment enters the "second hand" market, smaller farmers also will adopt these technologies. In the future, the questions remains if drones will be able to apply products "variable rate" and whether the safety and the risk of drift can be managed properly.

All in all, one may expect that aerial crop treatment ("crop dusting") will remain a service business. The technology and the markets are fairly mature: fixed wing platforms for large acre fields, drones an option on smaller fields/paddies. This leaves two basic observations. It is unlikely that fixed wing will be able to apply with variable rates. On the other hand, one may wonder if variable rate are useful on smaller fields.

When it comes to the areas of imagery from satellites, for fixed wing and lighter drone type platforms, it is obvious that the main issue remaining likely is not technical. Rather, who will make money from the services of capturing, analyzing and turning imagery data into actionable prescriptions, recommendations and actual applications in the field, the paddy and in the orchards? The study mentioned earlier (Erickson and Widmar) shows that in 2015 in the USA, only 1 of 5 retailers make profit from satellite imagery services. That figure came down from 1 of 3 in 2006. It is not obvious yet who in the USA makes profit from drone imagery services, except maybe specialized contractors.

The sharing of income from value created through imagery remains an issue in the USA.

Who will make money? Farmers? Distributors/Retailers? Players in the agricultural supply industry chemical/seed, fertilizer/nutrients, equipment, etc. ? Specialized service providers/ contractors? System integrators like IBM, Agrian, FBN, Climate Corp, some new, some active since a long time?

While US agriculture is sorting out this issue of business models, there are two very successful stories in the world, where this issue was resolved years ago. In France, medium size cereal farmers benefit from the collaboration, cost sharing and profit sharing between Spotimage (an aerospace company doing the images and information processing) and Arvalis (a technical support organizations paid for by the national organization of cereal growers doing ground truth and calibration, and the distribution/cooperatives doing the marketing, payment collection from farmers and agronomy recommendations). In Japan, small to medium wheat farmers in Hokkaido benefit from the cooperation between Japan Agriculture (JA) Cooperatives (marketing, collection, harvesting services and crop storage) and Hitachi (a technology company doing the images and the information processing).

What about China? China leads the world in production of light drones. China leads the world in applications by drone after having overtaken Japan. China is good at satellite imagery. China knows how to do application by airplane, both with fixed wing on large state farms, and with helicopter type platforms on smaller farms through contractors. China has an advanced fully integrated large farm management system developed by Nercita. Can China make it work profitably for all players involved? Probably yes, because it is recognized as a national priority.

One can expect that considering the size of the country and the wide diversity of agricultural situations, China can and will develop different technical solutions and business models of agricultural aviation for its different agricultural systems, and that will happen soon, before 2020. These diverse solutions will be copied by or offered/sold to other countries and agricultural situations.

作者简介

Marc Vanacht, Strategic Business Consultant. Research on Advanced technologies in agriculture, Precision Agriculture, Technology exchange, Bridge between markets and science, Technology & engineering, New product market identification, Development and market launch, and Product life cycle management. Early 2017, he was appointed by the King and the Government of Belgium and accepted by the USA Department of State as Honorary Consul of Belgium for the State of Missouri, based in St Louis.

植保无人机低空低容量喷雾的农药选择和制剂优化

闫晓静　孔　肖　袁会珠①

1　植保无人机应用现状

1.1　国外植保无人机发展概况

农用航空植保最早开始于 1911 年德国人提出用飞机喷洒农药控制森林害虫的计划。1949 年，美国开始研制专门用于农业的飞机，航空喷雾技术也有了很大的进步，从喷洒量大于 30 L/hm^2 的常规喷洒，发展到 5 ～ 30 L/hm^2 的低量喷洒，再到小于 5 L/hm^2 的超低量喷洒。1987 年，日本最先成功研制农用无人直升机，Yamaha 公司研制出载药量 20 kg 的喷药无人直升机 R-50。经过 20 多年的发展，日本目前已经拥有 2346 架农用无人机，成为世界上农用无人机喷药的第一大国，广泛应用于水稻、森林等病虫害防治。

美国是应用航空喷雾最发达的国家，也是目前世界上应用航空喷雾技术最为成熟的国家，具有较为完善的技术标准体系和先进严密的管理和技术支持研究机构。美国联邦航空局（FAA）颁布《联邦航空规章》[1]，制定的主要技术标准有《农业航空喷洒设备校准和分布模式测试》《测量地面、果园和空中喷雾飘移沉积过程的程序》等 20 多部[2]。美国农业部（USDA）对规范作业技术和作业质量进行管理，美国国土安全局（DHS）对作业设备的安全系统、喷施物的安全性等进行评估，美国农业航空协会（NAAA）、各州农业航空协会、航空喷施协会、有关大学与科研机构对航空喷洒作业技术进行研究[3]。

目前，发达国家对航空喷雾控制技术的研究热点主要集中在以下两个方面：①建立飞机喷雾的雾滴分布仿真数学模型，通过模型分析雾滴沉降规律，研究喷施高度、风速、不同飞机对雾滴粒径及雾滴飘移的影响。目前已投入使用的技术是可控雾滴技术，飞行员按作业条件选择相应的喷嘴和喷雾参数，达到控制雾滴种类、直径、飘移率等目的，取得最佳喷雾效果。②全球性定位系统（GPS）及精准施药技术在航空作业中的作用。进行航空植保作业时，通过控制 SSM 中不同区域（较小的面积单元）所需杀虫剂、肥料用量，进行变量喷施。

---

①　闫晓静、孔肖、袁会珠，中国农业科学院植物保护研究所。

1.2 国内植保无人机发展概况

我国航空植保始于20世纪50年代初期至60年代中期，1958年，南昌飞机制造厂生产的运－5投入使用，对我国农业航空的发展壮大做出了重大贡献。1963年，开始小麦病虫草害防治航空植保作业。到了20世纪90年代，出现了专门为轻型飞机如海燕等配套设计的农药喷洒设备，可广泛用于小麦、棉花等大田农作物的病虫害防治，化学除草、草原灭蝗、森林害虫防治以及喷洒植物生长调节剂、叶面施肥、棉花落叶剂等。[4]

近年来，我国无人机航空植保迅猛发展，无人机农业航空技术产品与材料的生产企业超过400多家，无人机航空植保作业在各地广泛应用。植保无人机从机型上可分为单旋翼无人机和多旋翼无人机，从动力上可分为油动和电动，从载重上可分为载荷5 L、10 L、12 L、15 L、16 L、18 L、20 L、25 L、30 L、40 L、45 L、58 L，70%的无人机载荷在10 L左右。

1.3 国内植保无人机施药技术发展现状

我国航空喷雾技术是在航空喷粉的基础上发展起来的，但随着航空喷雾的出现和对环境污染的重视，航空喷粉逐步被取代。航空喷雾可根据作业对象的不同，调控雾滴大小，且相比于粉剂，喷雾雾滴可在作物表面更好地沉积和附着，造成的农药飘移和对环境的污染也较少，因此，作业质量和防治效果较好。20世纪70年代以后，我国农业航空向低容量、超低容量和静电喷雾技术方面发展，同时，植保无人机航空喷洒系统、低空低容量喷洒、远程控制施药及低空变量喷药系统等农业航空喷洒基础理论研究也得到长足发展。[5-8]

随着植保无人机施药配套装备与技术的发展，植保无人机在农作物病虫害防治中高效、节水和节约劳动力的优势越发凸显。然而，在无人机进行植保作业的过程中，最突出的问题就是植保作业防效效果无法保证和作物药害问题，主要体现在：①药剂选择不当导致防治失败或出现药害；②剂型选择不当，造成喷头堵塞、药剂结块或者不同剂型药剂混合过程出现破乳结块，从而影响防治效果；③缺乏航空植保专用助剂或助剂添加不当，致使药剂蒸发、飘移，从而影响防治效果；④飞行速度、高度和喷幅等飞行参数选择不当而降低防治效果。因此，无人机植保作业药剂和制剂选择准则和助剂的评价标准是目前我国农业航空低空低容量喷雾技术所需解决的主要问题，也是制约我国无人机植保快速发展的技术瓶颈。

2 无人机植保作业药剂和制剂选择准则

2.1 无人机植保作业药剂的选择

我国农业病虫草害频发、重发和重大病虫害达120余种，年发生面积超过70亿亩次，造成潜在粮食损失3000亿斤。在目前相当长一段时间内，农药还是防治病虫草害的主要手段。无人机植保喷雾作业具有高浓度、飘移性和蒸发性等特点，所以应依据农药毒性分级和作用特性来选择。

2.1.1 农药毒性

农药毒性是指药剂进入生物体后，在短时间内引起的中毒现象。按照我国农药毒性分级标准，农药毒性分为剧毒、高毒、中等毒、低毒和微毒五级。

农药的安全性主要从对环境（土壤、水源和空气）、非靶标作物（蜂、鸟、鱼、蚕）、作物（敏感作物）、家畜（白鼠、兔子和山羊的健康毒理、急性、慢性和三致作用）、职业暴露（生产工人和施药农民）的安全性来综合评价。

农药的环境毒理主要通过蜂类急性毒性试验、鸟类毒性试验、鱼类毒性试验、家蚕毒性试验、藻类急性活性抑制试验、蚯蚓急性毒性试验、土壤微生物毒性试验、天敌赤眼蜂急性毒性试验、天敌两栖类急性毒性试验、非靶标植物影响试验、家畜短期饲喂毒性试验和大型甲壳类生物毒性试验等13个方面的试验来进行评价。

无人机植保作业药剂应选择低毒或微毒农药，同时兼顾其环境毒性，以便做出正确选择。例如，水稻田或者近水源的农田要特别注意药剂对水生生物的毒性。

2.1.2 农药雾滴的杀伤半径

植保无人机喷雾是低容量喷雾方式，农药雾滴在靶标上的沉积呈不连续性点状分布，一般每平方厘米农药雾滴在几个到几十个雾滴之间，因此，需要单个农药雾滴具有较大的“杀伤半径”，即一个雾滴具有较大的“杀伤面积”，才能保证对病虫害的防治效果。

为了明确药剂雾滴密度以及药剂浓度与作物病虫害防治效果之间的关系。笔者在室内通过Potter喷雾塔、离心式转盘雾化系统和ASS-3行走式喷雾塔模拟不同粒径的雾滴，分别选择吡虫啉、噻虫嗪、高效氯氰菊酯、阿维菌素和氟啶虫胺腈为供试药剂，选择小麦蚜虫为防治对象，研究不同雾滴粒径大小条件下杀虫剂浓度与其杀伤半径之间的关系，结果如图1至图3所示。

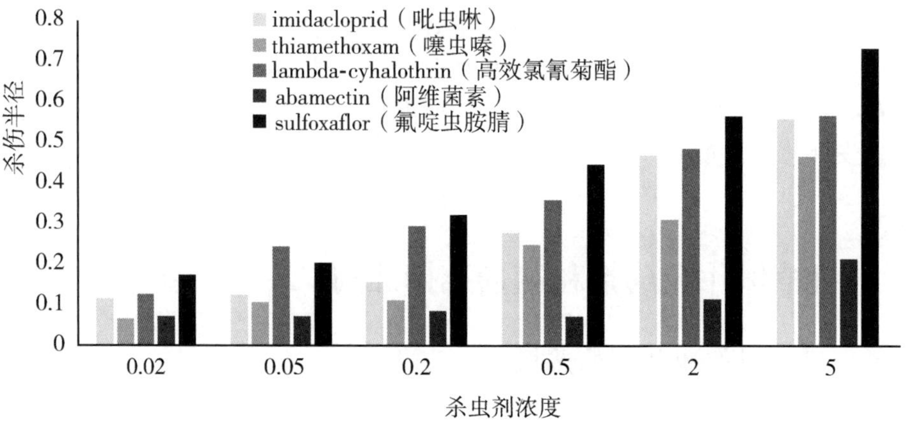

图1 Potter喷雾塔条件下杀虫剂浓度与杀伤半径之间的关系

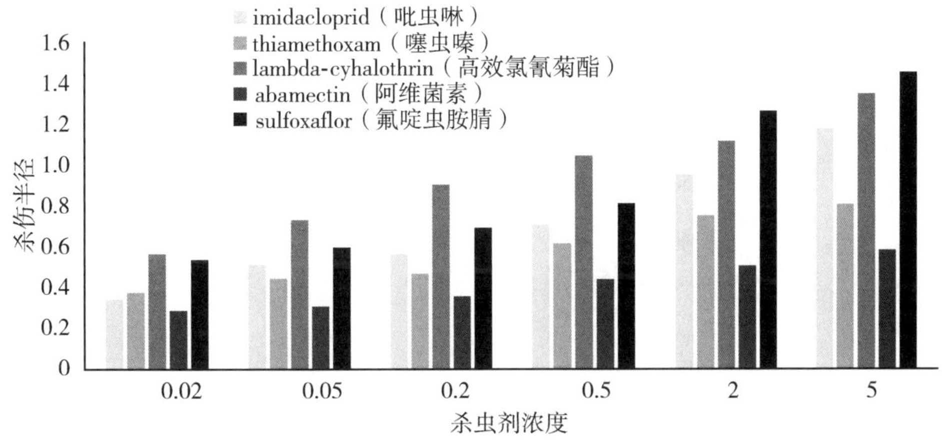

图 2　离心式转盘雾化系统条件下杀虫剂浓度与杀伤半径之间的关系

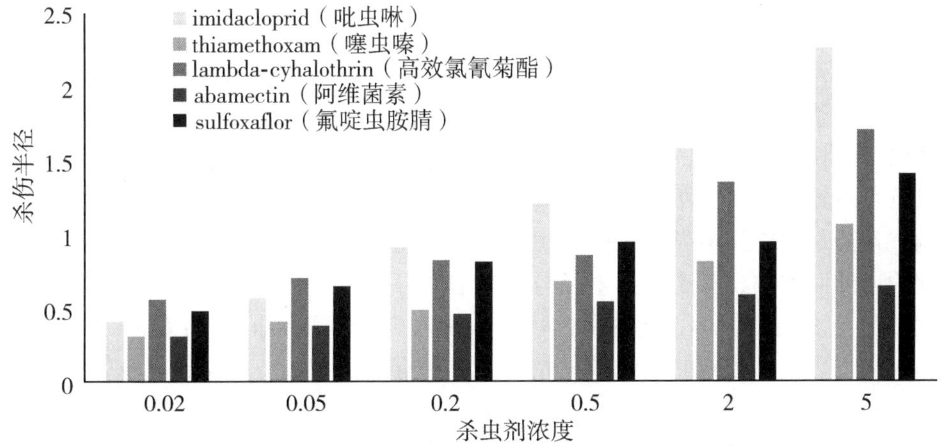

图 3　ASS-3 行走式喷雾塔条件下杀虫剂浓度与杀伤半径之间关系

表 1　Potter 喷雾塔模拟条件下不同杀虫剂的杀伤密度、杀伤半径和杀伤面积

	杀虫剂浓度	LN_{50} 喷雾粒径/cm^2	杀伤面积/mm^2	杀伤半径/mm
Potter 喷雾塔（43μm）	氟啶虫胺腈 sulfoxaflor 0.5g/L	81.8	0.61	0.44
	吡虫啉 imidacloprid 0.5g/L	214.0	0.23	0.27
	噻虫嗪 thiamethoxam 0.5g/L	283.6	0.18	0.24
	阿维菌素 abamectin 0.5g/L	2999.8	0.017	0.073

表 2　离心式转盘雾化系统条件下不同杀虫剂的杀伤密度、杀伤半径和杀伤面积

	杀虫剂浓度	LN_{50} 喷雾粒径/cm^2	杀伤面积/mm^2	杀伤半径/mm
离心式转盘雾化系统（153μm）	氟啶虫胺腈 sulfoxaflor 0. 5g/L	25. 1	2. 00	0. 80
	吡虫啉 imidacloprid 0. 5g/L	33. 2	1. 51	0. 69
	噻虫嗪 thiamethoxam 0. 5g/L	43. 2	1. 16	0. 61
	阿维菌素 abamectin 0. 5g/L	80. 7	0. 62	0. 44

表 1 和表 2 的研究结果显示：①在相同浓度、雾滴大小的情况下，不同药剂雾滴的杀伤半径、杀伤面积以及 LN_{50} 值不同。吡虫啉、噻虫嗪、高效氯氟氰菊酯以及氟啶虫胺腈对麦蚜的杀伤半径以及杀伤面积较大，适宜田间防治小麦蚜虫，而阿维菌素雾滴杀伤半径以及杀伤面积过小，LN_{50}值过大，不适宜田间喷雾防治小麦蚜虫。②在雾滴杀伤半径以及麦蚜致死率研究中，密度 LN_{50}受到雾滴粒径、药剂浓度以及雾滴密度三个因素影响，且药剂的杀伤半径、杀伤面积、麦蚜的死亡率随着雾滴粒径、药剂浓度以及雾滴密度的增加而增加，雾滴杀伤半径与药剂浓度一般符合对数或者线性拟合关系，致死中密度 LN_{50}与之相反。③分析不同粒径的雾滴对麦蚜产生相同防治效果时施药量情况的试验结果发现，在相同的 LN_{50}值的情况下，小雾滴（43 μm）施药量显著低于较大雾滴（153 μm，274 μm），并且呈现多倍关系。为此，田间喷雾防治麦蚜，在尽量避免小雾滴喷雾漂移、施药量相同的前提条件下，喷施小雾滴更有利于提高防治效果。

2. 2　无人机植保作业制剂的选择

适用于无人机植保作业的制剂应该是环境友好、抗蒸发、抗飘移且对作物安全的。地面常规喷雾技术的农药制剂要求悬浮率≥80%，分散率≥80%，粒径≤5 μm；无人机植保作业是一种低容量喷雾方式，对农药制剂要求更高，要求悬浮率≥95%，分散率≥95%，粒径≤2 μm。因此，常规制剂在无人机植保作业中受到极大冲击，发展航空超低容量专用剂型变为迫切需求。

从表 3 可以看出，目前适合于无人机植保作业的剂型为可溶液剂（SL）、悬浮剂（SC）和乳油（EC），可湿性粉剂（WP）和水分散粒剂（WDG）不适用于无人机植保作业。

表3 不同剂型药剂在无人机植保作业中的筛选比较

有效成分	制剂类型	溶解性	20 分钟内分层性	10 次飞行中堵喷头次数	喷雾粒径（μm）
吡虫啉	10% WP	有白色沉淀物	不分层	3	112.6
吡虫啉	70% WDG	有白色沉淀物	分层	2	121.8
吡虫啉	5% EC	乳化性高	不分层	0	76.1
吡虫啉	20% SL	乳化性高	不分层	0	78.6
己唑醇	5% SC	乳化性高	不分层	0	96.5
己唑醇	50% WDG	有灰色沉淀物	分层	3	106.4
吡蚜酮	25% WP	有灰色沉淀物	分层	3	137.1
吡蚜酮	25% SC	分散性高	不分层	0	94.8
吡蚜酮	50% WDG	1 分钟内不溶解	分层	2	136.5

3 无人机植保作业助剂评价标准

为满足植保无人机低容量喷雾技术的需要，需要添加喷雾助剂以防止雾滴飘移、增加雾滴润湿铺展性、提高雾滴沉积率，进而提高病虫害防治效果。本研究组主要从添加助剂前后理化性质（表明张力、润湿性、接触角和雾滴粒径）的改变、助剂的抗蒸发性和抗飘移来综合评价助剂，从而选择适宜的助剂种类和添加剂量。

3.1 助剂对药剂理化性质的影响

助剂最主要的作用就是通过改变药剂的理化性质，主要包括表面张力、润湿性、接触角和雾滴粒径来影响喷雾质量。分别测定在清水中添加不同剂量助剂（羟丙基瓜尔胶、甘油、Silwet DRS60、Greenwet360、Greenwet720）后其表面张力、润湿性、接触角和雾滴粒径的变化。助剂在清水中的添加量分别为 0.05%、0.10%、0.50%，测得助剂溶液的表面张力值见表 4。

从表 4 可以看出：① 0.1% Greenwet360 和 0.1% DRS60 能明显降液体的低表面张力，分别使清水的表面张力降低了 52.12 mN/m 和 50.96 mN/m。② 0.1% Greenwet360 的润湿性最好，0.1% DRS60 的润湿性次之。Greenwet360 和 DRS60 在药液润湿性测试卡上的润湿面积随浓度的增大而增大；助剂溶液的表面张力与润湿性呈正相关。③ DRS60能显著降低其在水稻叶面上的接触角。随着添加浓度的增大，在水稻叶面的接触角逐渐减小；当 Greenwet360 添加浓度为 0.50% 时，显著增大了其在水稻叶面的润湿性，可以在水稻叶面完全铺展开。④使用离心喷头在 0.3Mpa 压力下喷雾，添加 0.5% DRS60 对雾滴粒径影响最大，D_{V50} 为 165.59μm，比清水增大了 45.34μm；Greenwet360 对雾滴体积中径的影响程度次之，添加 0.5% Greenwet360 后雾滴体积中径达 148.54μm，比清水增大了 28.29μm。

表 4　不同助剂对药剂理化性质的影响

助剂	添加比例（%）	表面张力（mN/m）	润湿面积（个）	接触角（°）	D_{V50}（μm）
羟丙基瓜尔胶	0.05	54.84（±0.69）a	1	133.99	124.85
	0.10	49.14（±0.21）b	1	131.30	119.21
	0.50	47.32（±0.42）c	2	133.20	122.20
甘油	0.05	71.33（±0.08）a	1	121.00	119.34
	0.10	71.48（±0.06）a	1	129.23	115.77
	0.50	71.47（±0.06）a	1	119.77	126.45
DRS60	0.05	34.90（±0.08）a	1	75.42	120.69
	0.10	26.79（±0.21）b	2	37.56	135.92
	0.50	22.58（±0.24）c	5	25.73	165.59
Greenwet360	0.05	34.08（±0.19）a	1	122.94	131.03
	0.10	28.08（±0.07）b	2	112.62	131.31
	0.50	22.28（±0.12）c	4	36.76	148.54
Greenwet720	0.05	35.1（±0.11）b	4	93.35	122.53
	0.10	36.03（±0.11）a	4	95.07	126.90
	0.50	35.02（±0.38）b	4	117.11	116.36
清水		72.19（±0.42）	1	129.34	120.25

注：表中同列数据后标相同小写字母者表示经 Duncan 氏新复极差法检验，在 P0.05 水平上差异不显著。

3.2　助剂的抗蒸发性

无人机植保作业易受环境因素的影响而在施药过程中发生蒸发飘移损耗，影响对病虫害的防治效果。因此，助剂的抗蒸发性能对无人机植保作业效果具有重要影响。本文通过利用视频光学接触角测量仪 OCA20 测定不同种类助剂在田间施药平均温度（33℃）下对雾滴蒸发的影响，以评价助剂的抗蒸发性能，结果见表 5。该结果表明，在 33℃时，不同浓度的五种助剂均不同程度地降低了雾滴的蒸发速率。其中，0.05% DRS60、0.10% DRS60 对雾滴蒸发的抑制效果最好。

表 5　添加助剂对雾滴蒸发体积的影响（33℃）

添加助剂	添加比例（%）	雾滴蒸发方程	相关系数
清水		y = -0.0024x + 2.7695	0.9863
DRS60	0.05%	y = -0.0013x + 2.6662	0.9659
	0.10%	y = -0.0022x + 2.6853	0.9675
	0.50%	y = -0.0021x + 2.65	0.9889
	1.00%	y = -0.0019x + 3.2152	0.9837
Greenwet360	0.05%	y = -0.0013x + 3.5021	0.941
	0.10%	y = -0.0013x + 3.1277	0.9593
	0.50%	y = -0.0018x + 3.0691	0.9875
	1.00%	y = -0.002x + 3.4239	0.9883
Greenwet720	0.05%	y = -0.0019x + 3.6023	0.9807
	0.10%	y = -0.0017x + 3.6204	0.944
	0.50%	y = -0.0018x + 2.8732	0.9676
	1.00%	y = -0.0017x + 3.7259	0.9765
甘油	0.05%	y = -0.0016x + 2.8592	0.9851
	0.10%	y = -0.0018x + 3.1654	0.9742
	0.50%	y = -0.0024x + 3.1435	0.9598
	1.00%	y = -0.0019x + 3.7957	0.9797
羟丙基瓜尔胶	0.05%	y = -0.0019x + 3.4315	0.9867
	0.10%	y = -0.0017x + 3.7009	0.9442
	0.50%	y = -0.0013x + 3.7181	0.9099
	1.00%	y = -0.0014x + 3.5711	0.9652

3.3　助剂的抗飘移性

植保无人机风场造成的药剂飘移是影响无人机植保作业效果和环境安全的重要因素。因此，助剂的抗飘移性能也是助剂评价的又一重要指标。本文在风洞条件下测定不同种类助剂对雾滴密度和沉积量的影响，以评价助剂的抗飘移性能。

由图 4 和图 5 可知，随着测定点距离喷头位置的增大，各助剂处理在地面的雾滴沉积量逐渐减少。测定点距离喷头位置为 0.5 m 时，清水、DRS60、Agrospread910、NIS、Greenwet360、Greenwet720 的雾滴沉积量分别为 0.94 μg/cm²、1.63 μg/cm²、1.66 μg/cm²、

1.38 μg/cm²、1.10 μg/cm²、0.32 μg/cm²。测定点距离喷头位置为3m时，助剂的雾滴沉积量分别为0.36 μg/cm²、0.27 μg/cm²、0.33 μg/cm²、0.36 μg/cm²、0.33 μg/cm²、0.14 μg/cm²。这表明，当添加浓度为0.50%时，DRS60的抗漂移性能最好，Agrospread910次之。

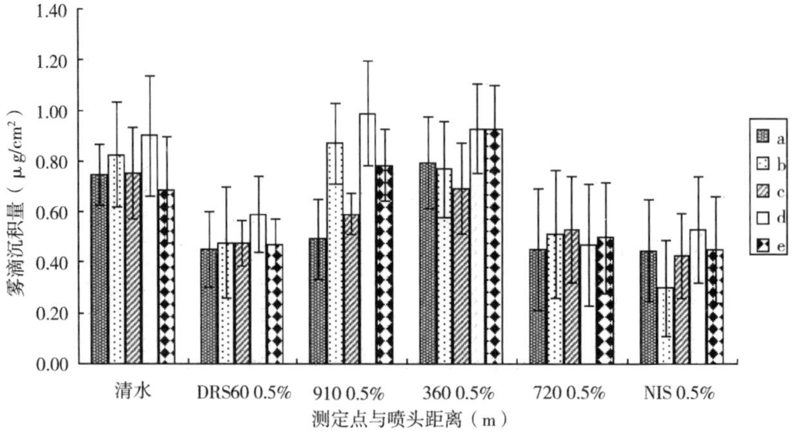

图4　不同助剂对雾滴沉积量的影响（平面雾滴采集器）

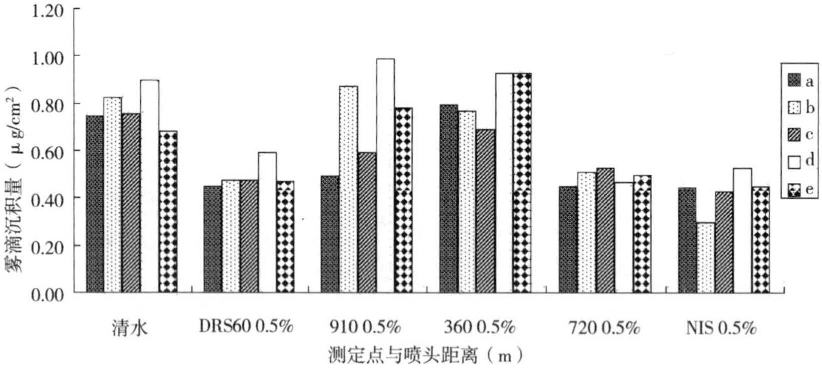

图5　不同助剂对雾滴沉积量的影响（线状雾滴采集器）

4　结论

药剂与植保无人机堪比子弹与枪，两者之间是相辅相成、不可或缺的两部分。因此，在无人机植保作业过程中，使用药剂应遵循低毒环保、低剂量高杀伤半径的准则，制剂以航空超低容量专用制剂为发展方向，同时兼顾助剂在传统剂型中发挥的作用。以表面张力、润湿性、接触角、雾滴粒径、抗蒸发性和抗飘移性几个方面来全面评价助剂的性能。

相信随着科技的不断发展，我国会陆续出现一批航空专用助剂和航空超低容量专用制剂，从而真正让植保无人机拥有健硕的双翼，飞驰在我国广袤的农田，大幅提高病虫害防治效率，促进我国现代农业的可持续快速发展。

参考文献

[1] KIRK I W. Measurement and predication of helicopter spray nozzle atomization [J]. Transactions of the American society of agricultural engineers, 2002, 45 (1): 27-37.

[2] YUBIN L, STEVEN J T, YANBO H, et al. Current status and future direction of future direction of precision aerial application for sitespecific management in the USA [J]. Computers and electronics in agriculture, 2010, 74: 34-38.

[3] 薛新宇，兰玉彬. 美国农业航空技术现状和发展趋势分析 [J]. 农业机械学报，2013，44 (5): 194-202.

[4] 周志艳，臧英，罗锡文，等. 中国农业航空植保产业技术创新发展战略 [J]. 农业工程学报，2013，29 (24): 1-10.

[5] 王军，简捷，杨道训，等，农业航空喷雾雾滴飘移及其数学分析 [J]. 植物保护学报，1994，21 (3): 275-281.

[6] 茹煜，贾志成，范庆妮，等. 无人直升机远程控制喷雾系统 [J]. 农业机械学报，2012，43 (6): 47-52.

[7] 王玲，兰玉彬. 微型无人机低空低容量喷药系统设计与雾滴沉积规律研究 [J]. 农业机械学报，2016，47 (1): 15-22.

[8] 秦维彩，薛新宇，周立新，等. 无人直升机喷雾参数对玉米冠层雾滴沉积分布的影响 [J]. 农业工程学报，2014，30 (5): 50-56.

作者简介

袁会珠，1967 年 1 月出生，河北藁城人。从 1992 年开始，历任中国农业科学院植物保护研究所研实员、助理研究员、副研究员、研究员，博士生导师，农业农村部农药研制与施药技术重点实验室学术委员会主任。长期从事农药使用技术研究工作，承担了国家自然科学基金、973 计划项目、863 计划项目、公益性（农业）行业科研专项等项目，在新农药筛选与作用机理、种衣剂和省力化农药制剂、农药喷雾检测、植保机械与施药技术等方面都有比较深入的研究。出版了《现代农药应用技术图解》《农药应用指南》《农药使用技术指南》《农药安全使用知识》《植保机械与施药技术规范化》等书籍。发表文章 200 余篇。2015 年，被中国农业工程学会农业航空分会授予“农业航空发展贡献奖”。

Airborne Remote Sensing Using a Flexible Sensor System for Gyrocopter

Jens Bongartz Caspar Kneer Alexander Jenal Immanuel Weber[①]

1 Introduction

Using UAVs will bring a dramatic change in airborne remote sensing and plant protection. But therefore autonomous beyond visual line of sight (BVLOS) operations with high payloads are necessary. We estimate that this technology will take at least five years to mature. To fill this time gap, we actually focus on manned gyrocopter technology, which we present in this paper. Gyrocopter is a type of rotorcraft that uses an unpowered rotor in autorotation to develop lift and an engine-powered propeller to provide thrust. They are easy and inexpensive to operate and their flight characteristics are suitable for remote sensing.

Furthermore, gyrocopter is well suited for large-scale surveying and as a platform for experimental sensor setups. Gyrocopter can carry up to 100 kg payload and cover up to 2,000 hectares per hour by an endurance of 3 to 4 hours flying time. Beyond the gyrocopter we present different sensor-systems we have developed and adapted so far.

2 The Gyrocopter

Unmanned Aerial Vehicles (UAV) and Remotely Piloted Aerial Systems (RPAS) are emerging trends at the lower end of investments, payload and endurance as well as areal coverage per hour. Moreover, these devices need special flying permissions and have to stay always in line of sight, so it reduces their reach of action drastically. In contrast, a manned Gyrocopter is a full-fledged participant of the general air traffic. With its ability to carry sensor payloads up to 100 kg for flight times up to five hours a gyrocopter can perfectly fill the gap between UAVs and satellite missions. Combined with a low acquisition price and maintenance costs a gyro is ideally suited as an agricultural remote sensing platform.

① Application Center for Multimodal und Airborne Sensing (AMLS), Remagen / Germany Fraunhofer Institute for High Frequency Physics and Radar Technique (FHR) University of Applied Sciences Koblenz.

2.1 Auto-Gyro MTOsport

We developed a sensor system specifically designed for the gyrocopter model MTOsport of the manufacturer Auto-Gyro from Hildesheim, Germany (see Fig. 1). This gyrocopter is the most widespread gyrocopter with more than 1,500 units sold worldwide. This tandem seater is a very reliable and robust aircraft. Auto-Gyro has a subsidiary company in China (Auto-Gyro China) and is establishing a distributor and maintenance network in China. Auto-Gyro is actually the only gyrocopter manufacturer certified by the Civil Aviation Administration of China (CAAC).

Figure 1　Auto-Gyro MTOsport equipped with sensor-equipment

3 The Sensor-System

The main system is segmented in three subsystems. The first part is a base system that provides aircraft independent battery power and position data from a high-precision GNSS. Data from this part are forwarded to the two subsystems described below.

The second part is the flight management system (FMS). It is essential for navigating the aircraft to a preplanned area and emitting a trigger signal for the sensor system at pre-calculated trigger points. The third part consists of the sensor management system (SMS) as well as the custom-developed and integrated sensors. The SMS controls the particular sensors and acquires the data along with further information about position, flight level, incident light, and environmental data at a given trigger impulse by the FMS. The tailor-made control and data acquisition application hosted on the SMS is based on our own software framework called AWS. Once configured and started, the overall system runs autonomously.

3.1 FlugKit

We developed a mechanical base system, which can be installed in every MTOsport on the rear-seat, which is called FlugKit (see Fig. 2). The idea behind this system is that we can use locally available gyrocopter and we do not have to bring our own gyro to the survey area.

Furthermore, a gyrocopter can be used more versatilely. For the installation of the FlugKit, no modifications to the aircraft are necessary. Hence, when the FlugKit is removed, the gyrocopter can be used for passenger flights or can be equipped with spraying devices. Installation and removal of the FlugKit takes around 30 minutes.

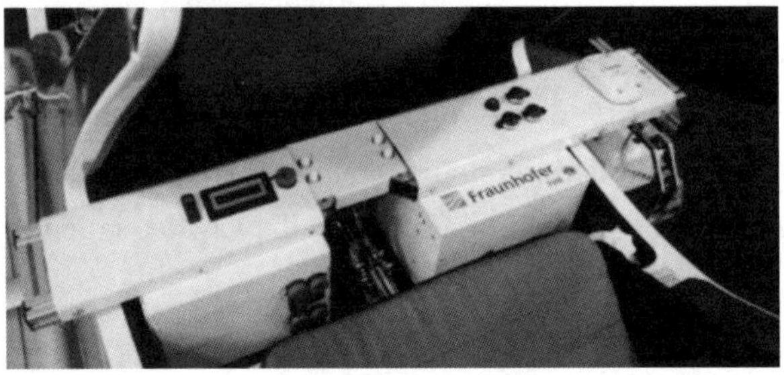

Figure 2 FlugKit installed in the rear-seat of a MTOsport gyrocopter

The FlugKit is attached mechanically via a clamp to the main frame of the gyrocopter and consists of two containers with power supply, embedded computer and GNSS receiver. Outside of the cabin is a gimbal, which can host different cameras and sensors as described below.

3.2 High-Resolution RGB-Imaging

A high resolution (36 MP) DSLR Nikon D800e is used to acquire aerial VIS-RGB images. The lens has a fixed focal length of 35mm and is very light sensitive. At a typical flight level of 400 m AGL the swath width is 411 m resulting in a ground resolution of 0.06 m. A custom developed hardware connects the camera with the GNSS-system, the FMS as well as the SMS. Therefore a specific image acquisition by a trigger signal is possible and the images are directly georeferenced in meta-data and then stored on the camera. A further integration concerning storing the image data into the SMS is planned.

3.3 Hyperspectral Imaging (PanHyper)

The second sensor modality is a hyperspectral camera (based on a Cubert UHD 285 system [1]). The advantage of such a system is the direct acquisition of areal hyperspectral image information that eliminates the need of an expensive, high precision IMU that is necessary for image reconstruction from line scanner data. The spectral range of the camera is 450 nm to 950 nm. The spatial resolution amounts to 2,500 (50 × 50) spectral pixel. Each pixel consists of 125 spectral channels with 14-bit dig it ization. The integrated panchromatic reference channel has a spatial resolution of 1,000 × 1,000 pixel. At flying altitudes of 300 m AGL the swath width is about 100 m with a h yperspect ral ground resolution of 2 m per pixel.

3.4 Thermal Imaging (PanTIR)

The PanTIR imager is a thermal imaging system. To overcome the usual drawback of the low spatial resolution of aerial thermal infrared (TIR) images, resulting from low sensor resolution (here: 640 × 480 pixel) a dual camera setup was developed. The thermal information is acquired by an uncooled microbolometer in the long-wavelength infrared range (LWIR). As a second modality for acquiring high-resolution, well-structured image data a panchromatic VIS camera (4 MP, 12 bit) was selected. Both sensors are precisely aligned in a custom designed rack that is mounted in a stabilization system (gimbal). After a survey flight the acquired data sets of the two sensors are fused to multi-channel images using the workflow in Weber, etc. [2] The resulting images can be processed much easier with photogrammetric tools as the panchromatic channel is used for image alignment and processing (see Fig. 3). The resulting transformations can then be applied to the thermal channel.

Figure 3 Thermographic aerial map: 24 km^2 were captured within 45 minutes

3.5 Multispectral Imaging (PanX)

The latest sensor development is based on the above-mentioned PanTIR imager. However, instead of an LWIR camera up to five different VIS-NIR can be combined with the already known VIS-PAN camera model. Each NIR-enhanced camera can be equipped with a custom made filter in order to select the requested spectral bands. For precise alignment of the cameras a special rack was designed. A system with four channels was created and tested with a camera setup for NDVI calculations. [3] Therefore, besides the panchromatic channel, a red channel (670 nm ± 5 nm) and a near-infrared channel (850 nm ± 20 nm) were chosen. Each camera has a resolution of 4 MP. At a flying altitude of 300 m AGL the swath width is about 300 m with a ground resolution of 0.15 m per pixel (see Fig. 4).

Figure 4　Multispectral aerial map with calculated NDVI-GSD: 15 cm

4　Conclusion

We presented a remote sensing system, FlugKit, which is a modular container system that contains an independent power supply unit, a precision navigation system, an embedded computer system for data-acquisition and storage plus the sensors. A multi- and hyperspectral camera system for the VIS-NIR spectral range (PanX and PanHyper) and a thermal infrared system (PanTIR) are available as sensors as well as a high-resolution digital RGB camera. Additional sensor modalities are under development or can be developed on demand. One version of the FlugKit is specially designed for the gyrocopter model Auto-Gyro MTOsport, which is the most widespread gyrocopter in the world. The MTOsport is certified and available on the Chinese market. The FlugKit can be installed into and removed from any available MTOsport within 30 minutes. No modifications of the aircraft are necessary.

References

[1] MICHELS R, LIEBSCH S, GRASER R. Snapshot hyperspektroskopie[J]. Photonik, 2014(1): 36-38.

[2] WEBER I, JENAL A, KNEER C, et al. PanTIR A dual camera setup for precise georeferencing and mosaicing of thermal aerial images[C]. Munich: Proceedings on workshop of photogrammetric image analysis, 2015.

[3] THENKABAIL P S, SMITH R B, DE PAUW E. Evaluation of narrowband and broadband vegetation indices for determining optimal hyperspectral wavebands for agricultural crop characterization[J]. Photogrammetric engineering and remote sensing, 2002, 68 (6): 607-622.

作者简介

Dr. Jens Bongartz, Professor, Vice-President for Research University of Applied Sciences Koblenz from 2008 to 2012. Head of the Application Center for Multimodal and Airborne Sensing (AMLS), Fraunhofer Institute for High-Frequency Physics and Radar Technology (FHR) now. Research and Teaching into Sensors and Signal Processing.

大型有人驾驶飞机精准施药系统及施药质量评价技术研究

陈立平　张瑞瑞　唐　青　徐　刚①

1　引言

航空施药过程中的农药飘移是造成相关环境问题的根源，其可能破坏生态资源、危害非目标区域作物及牲畜以及对人员健康造成不良影响。[1-4]因此，如何进行精准施药，减少药液损失、提高农药利用效率、降低环境次生危害是农业航空精准施药技术的研究目标。

在国家政策、科研项目以及应用市场的推动下，我国农业航空施药技术得到了快速发展，在农用航空机型、航空施药关键技术、航空施药配套装备等方面取得了系列成果。[5]但是，与发达国家相比，我国在航空施药基础理论和核心体系化技术方面，特别是低空喷雾沉降规律、大型飞机变量施药控制技术、雾滴沉积检测技术等方面还存在较大差距。[6]

本文主要介绍本团队在航空精准施药实现及质量评价技术上取得的进展，包括航空施药导航控制及变量施药系统的开发及取得的应用效果，航空施药雾滴飘移沉积预测评价模型的构建，雾滴沉积质量评估的新方法。

2　航空施药导航控制及变量施药系统

飞行员驾驶有人飞机进行植保作业时，飞机作业路径依赖于飞行员的视线及个人经验，很难保证作业区域的全覆盖，常常出现重喷、漏喷等情况。[7]该系统主要包含机载作业导航系统和机载变量施药控制系统，以施药作业导航为核心，在硬件设备采集到的实时定位信息的基础上，面向有人驾驶飞机配置航空施药导航功能，以适应飞行作业中的严苛作业环境。[8]

①　陈立平、张瑞瑞、唐青、徐刚，北京农业智能装备技术研究中心，北京市农林科学院，国家农业智能装备工程技术研究中心，国家农业航空应用技术国际联合研究中心。

2.1 机载航空施药导航系统

机载航空施药导航系统包括三个核心功能：施药区域作业规划、施药路线导航和施药质量实时计量评估。实验和应用结果表明，该系统路径规划效率高，施药作业时地面平均覆盖率可达96%，可以满足有人机施药作业及路线规划需求。系统可以应用于具有复杂边界或具有大量小块农田的区域。在2013年至2016年，该系统已广泛应用于Bell 407、R66、Air Tractor 802F等主流农用有人驾驶固定翼飞机和直升机，服务作业面积超过300万亩。

选取10种典型农业航空施药作业任务，对系统的使用该导航系统后的遗漏面积、施药覆盖率等指标进行评估。（见表1）在使用该系统后，所有作业任务施药平均覆盖率大于96%，相较于不使用该系统时，施药遗漏率大幅减少，施药覆盖率显著提升。

表1　使用机载航空施药导航系统后的总遗漏面积和施药覆盖率

任务编号	总施药面积（公顷）	总遗漏面积（公顷）	施药覆盖率（%）
1	113.5	3.4	97.1
2	169.5	2.1	98.8
3	184.8	6.5	96.6
4	205.0	7.8	96.3
5	247.5	5.2	97.9
6	292.7	11.6	96.2
7	312.1	11.3	96.5
8	325.0	6.2	98.1
9	383.9	7.2	98.2
10	429.4	17.8	96.0

2.2 机载变量施药控制系统

机载变量施药控制系统通过在药泵和喷头间加入流量控制阀门、流量检测传感器，并综合飞机飞行速度、作业幅宽等信息，实现飞机作业过程施药量的精准自动控制。使用该系统后，在飞机施药过程中，系统会根据飞机当前飞行速度自动调整管道流量，将单位面积施药量严格控制在用户设定值，实现均匀施药。[9] 机载变量施药控制系统实物如图1所示。

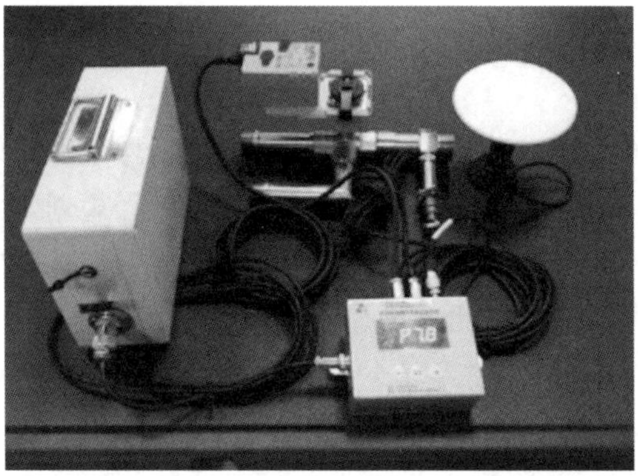

图1　机载变量施药控制系统实物

通过测量变量施药控制系统在不同速度情况下实际施药量的变化情况，来验证系统的实际性能。实验设定作业幅宽为 30 m，施药量为 6.0 L/hm^2，然后分别测量速度为 90 km/h、110 km/h、120 km/h、160 km/h 条件下实际施药量的变化情况。实验过程中，为了保证实验结果的严谨性与可靠性，同一速度下重复测量 3 次，取 3 次测量结果的平均值作为有效值，实验过程中数据的记录频率为 1 Hz。

表2　流量控制实验数据

速度（km/h）	设定施药量（L/hm^2）	实际施药量（L/hm^2）	相对偏差（%）
90	6.0	6.3	5.0
110	6.0	5.6	6.7
120	6.0	6.2	3.3
160	6.0	6.5	8.3

从表 2 中可见在不同实验速度下测量得到的设定施药量、实际施药量及二者之间的相对偏差。从表 2 所示实验结果可以看出，当速度在 90～160 km/h 范围内时，测量得到的实际施药量与设定施药量的最小偏差为 3.3%，最大偏差为 8.3%，即设定施药量和实际施药量的偏差范围始终保持在 10% 以内，有效抵消了飞机速度波动引起的施药量变化，提升了施药量的均匀性。

3　航空施药雾滴飘移沉积预测模型

雾滴地面沉积分布和飘移距离是航空精准施药药效评价的重要指标之一。而雾滴的飘移运动主要受环境风场的影响，包括大气边界层流动和飞机尾流，其流动十分复杂，难以预测。[10-16] 针对这一问题，本团队构建了大气边界层和飞机尾涡的计算流体力学模型，以田间实验测量得到的大气边界层速度剖面信息作为边界条件，通过数值模拟的方

法计算飞机尾涡在大气边界层中的运动，研究影响雾滴运动的风场特征，预测雾滴飘移沉积。[17,18]

3.1 模型分区及控制方程

模型首先根据典型流动状态对计算域进行分区，在不同区域采用不同的网格密度，在保证模拟精度的基础上减小计算量。根据流动特点将流场划分为不同的区域（见图2），主要包括欧拉区、雷诺应力区和大涡模拟区。区域的划分并不以流动控制方程为依据，而是以网格间距为主要因素。欧拉区主要集中在远场范围，区域剪切小，流动平滑；雷诺应力区主要位于壁面附近，区域要求较高的网格密度，以准确应用壁面函数进行区域内的物理量计算；大涡模拟区主要涵盖尾涡、二次涡等区域，通过对流场进行滤波，过滤小尺度涡的影响，从而在保证模拟精度的同时极大地减小计算量。

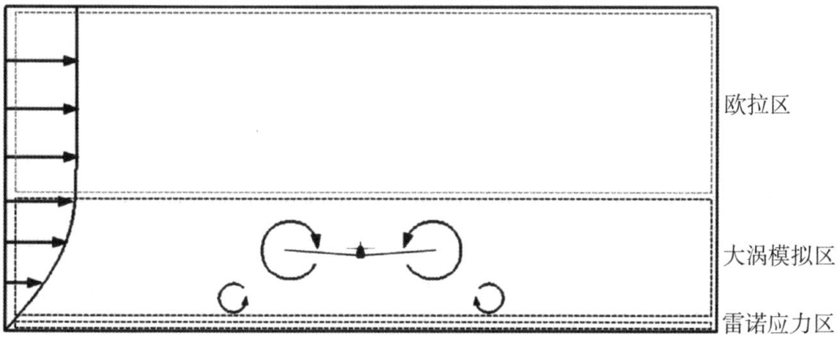

图2　网格流动区域划分

以N-S方程中的质量守恒方程、动量守恒方程和能量守恒方程为基础，通用数学描述形式为

$$\frac{\partial(\rho\varphi)}{\partial t}+\operatorname{div}(\rho u\varphi)=\operatorname{div}(\Gamma_{\varphi}\ \mathrm{grad}_{\varphi})-\frac{\partial\rho}{\partial x}+S_{\varphi}。$$

式中，φ为通用变量，代表u、v、w、T等求解变量；Γ_{φ} 为广义扩散系数，S_{φ} 为广义源项。

3.2 模型边界条件设置

基于田间实验测量大气边界层速度剖面，通过在不同高度布置三维超声风速传感器，并使用多通道数据采集卡同步采集测量数据。实验获得的数据用于CFD模型的侧风边界条件。三维超声风速传感器田间布置及安装方式如图3所示。

图 3　三维超声风速传感器阵列布置及安装方式

为了精确测量典型作业条件下的大气边界层平均速度剖面及脉动速度剖面，实验风速采集频率达 20 Hz。基于田间实验测得的风速数据，利用 k-ε 模型构建速度入口条件如下所示。

$$U(z)=\frac{u_*}{\kappa}ln\left(\frac{z+z_0}{z_0}\right)$$

$$k(z)=\frac{u_*^2}{\sqrt{C_\mu}}$$

$$\varepsilon(z)=\frac{u_*^3}{\kappa(z+z_0)}$$

用田间实测结果换算 k、ε 值，并以换算值作为参考，通过摩擦速度 u_* 的选取修正速度入口条件。其中，$z_0=0.06$ m 为地表粗糙度特征尺度；$C_\mu=0.0287$，Karman 常数 $\kappa=0.4187$；摩擦速度 $u_*=0.406$ m/s。上述三式得到的结果与飞行实验测量结果对比如图 4 所示。

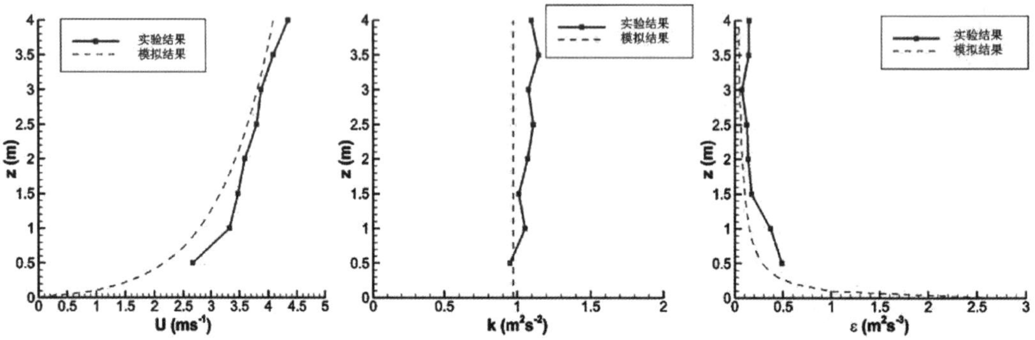

图 4　平均速度剖面 U、湍动能 k 及湍流耗散率 ε

3.3　模型雾滴注入初始条件设置

采用 Malvern 激光衍射系统在国家农业智能装备工程技术研究中心 IEA-I 型高速风洞中测量 AU 5000 转笼雾化器产生的雾滴粒径分布，粒径分布数据用于 CFD 模型初始条件。转笼雾化器实验平台如图 5 所示。

图 5　转笼雾化器雾滴粒径参数测量

实验中，转笼雾化器叶片角度分别设定 45 deg、55 deg、65 deg、75 deg 四个变量；流量设定 10 L/min、15 L/min、20 L/min、25 L/min 四个变量；风速设定 185.5 km/h、218.4 km/h、253.5 km/h 三个变量，测点距离雾化器 460 mm。实验测得的 AU 5000 型转笼雾化器雾滴粒径分布如表 3 所示。

表 3　AU 5000 型转笼雾化器雾滴粒径分布

流量 (L/min)	扇叶角 (deg)	风速 185.5 km/h			风速 218.4 km/h			风速 253.5 km/h		
		VMD (μm)	<100μm (%)	<200μm (%)	VMD (μm)	<100μm (%)	<200μm (%)	VMD (μm)	<100μm (%)	<200μm (%)
10	45	159.7	23.67	63.65	111.7	42.44	87.2	91.15	56.49	93.94
	55	184.9	15.08	55.87	140.7	27.53	75.23	109.2	43.88	87.92
	65	220.6	10.66	43.45	164.5	20.88	63.22	131.7	32.71	75.48
	75	223	10.13	42.24	170.4	19.23	61.16	134.6	30.96	76.65
15	45	170.8	19.16	60.69	128.2	33.97	77.95	100.4	49.73	91.2
	55	222.8	10.4	42.69	170.5	19.78	60.51	116.7	39.5	84.83
	65	216.2	10.22	44.21	165.9	19.3	64.09	132.3	31.38	78.98
	75	243.1	8.381	37.19	184.4	15.99	55.72	146.2	27.13	70.03
20	45	178.8	16.85	58.01	131.4	32.1	78.08	106.8	45.26	89.48
	55	208.2	11.94	47.19	160.4	21.12	65.85	125.1	35.07	80.92
	65	223.8	10.86	42.52	169.1	19.64	61.78	132.7	32.01	77.29
	75	248.9	7.731	35.36	186.3	14.81	55.24	147.3	20.46	71.83

（续表3）

流量（L/min）	扇叶角（deg）	风速 185.5 km/h			风速 218.4 km/h			风速 253.5 km/h		
		VMD（μm）	<100μm（%）	<200μm（%）	VMD（μm）	<100μm（%）	<200μm（%）	VMD（μm）	<100μm（%）	<200μm（%）
25	45	192.5	13.67	52.81	144.2	26.2	73.23	113.3	41.2	86.76
	55	224.1	9.158	41.8	166	19.15	63.84	133.2	30.36	78.72
	65	243.3	7.891	36.65	184.2	14.79	56.1	145.7	25.66	72.61
	75	261.3	6.494	32.3	197.1	13.18	51.04	157.3	21.88	67.2

从实验结果看，转笼整体载荷增大导致转速变慢，流量增大则会导致雾滴体积中径整体增加。叶片角度减小及风速增大导致雾滴体积中径减小，这一方面是叶片角度减小及风速增大导致转笼转速增加，另一方面是风速增大对雾滴造成的剪切破碎作用更强。

3.4 雾滴飘移沉积预测结果

转笼雾化器产生的雾滴粒径分布作为初始条件被通过用户自定义函数注入 CFD 模型中，之后在飞机尾涡风场及重力作用下开始产生飘移及沉积运动。

雾滴释放后不同时刻的空间分布及运动情况如图 6（a）所示：从 2 s 时的雾滴空间分布可以看到尾涡对雾滴运动的巨大影响。靠近尾涡的雾滴被迅速卷入尾涡的诱导气流中，围绕涡核形成两个环状喷雾带。其中，小雾滴更易受尾涡诱导速度的影响，留空时间较长。因蒸发作用，雾滴粒径进一步减小，导致更易产生飘移损失；而粒径较大的雾滴主要在水平诱导速度的影响下改变运动轨迹，沿展向外侧运动。两涡之间的雾滴则在下洗气流的作用下加速向地面沉降。10 s 时，大雾滴基本沉积到地面，而卷入尾涡的小雾滴在湍流脉动的影响下逐渐扩散，空间分布范围不断增大。与此同时，尾涡之间的距离增大，携带着雾滴向两侧运动，尾涡强度也不断衰减。

图 6（b）对比了 CFD 和 AGDISP 计算得到的雾滴在地面的沉积量：两种方法预测得到的喷幅内的雾滴沉积曲线十分接近，雾滴沉积规律相似，沉积量也基本相同。在下风方向的喷幅之外，雾滴沉积量迅速减小，两条曲线之间出现差异。AGDISP 沉积曲线上可以看到一个小的峰值，是由右侧尾涡携带喷雾旋转之后释放造成的。而在 CFD 模拟中，右侧涡强度迅速衰减，携带雾滴的能力下降，湍流脉动的扩散作用使雾滴分布趋于均匀，因而沉积曲线光滑，并未出现局部的峰值。

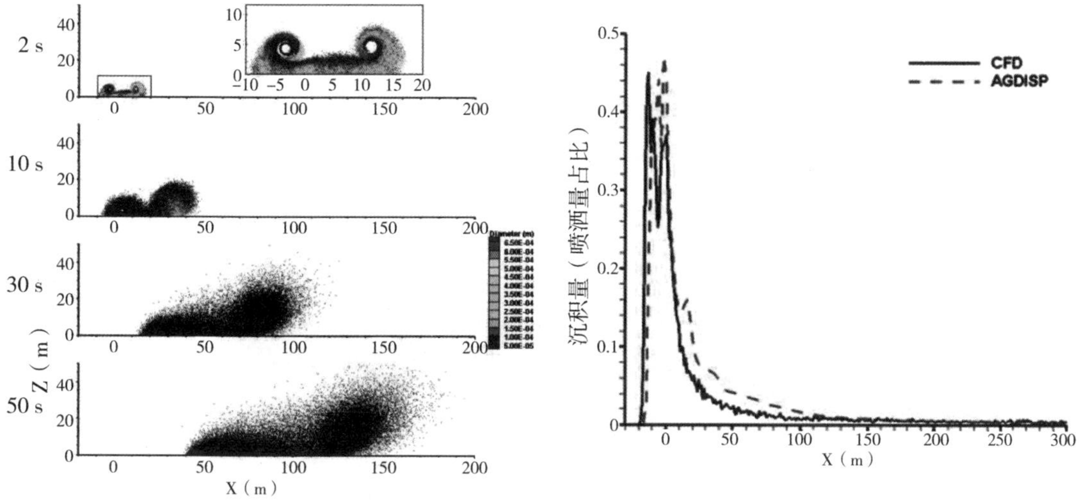

图6　不同时刻雾滴粒径分布及空间位置（a）和地面沉积量 CFD 模拟结果和 AGDISP 模拟结果对比（b）

4　雾滴沉积质量评估新方法

雾滴沉积质量与作业方式、喷雾参数、气象环境等密切相关。目前，国内外研究人员已对农业航空施药质量及其影响因素进行了大量研究工作。[19-29]一般来说，雾滴沉积质量主要从雾滴沉积密度、均匀性、飘移损失量等方面进行评估。[30-34]除了传统水敏纸、示踪剂等雾滴沉积质量评估方式，本团队还研究了快速、实时、可数字化的喷雾质量田间评价新方法。

4.1　基于红外热成像及可见光的雾滴沉积效果测量

在航空施药过程中，由于从喷嘴释放的液滴的温度和透射率不同于环境大气的温度和透射率，其红外热辐射的强度可以反映液滴浓度。因此，使用非制冷红外热成像系统和基于背景剔除的图像处理技术可实现对雾滴云团运动过程的监测。该方法具有非接触性、实时性、可测空间范围大等优势。[35]

实验检测现场如图 7 所示：Thrush 510G 飞机逆风进入实验场，当飞机到达标记点时，飞行员打开 AU 5000 雾化器并开始施药。实验使用 21 个水敏纸测试点。第 11 点位于标记位置，其他点均匀分布在标记点的两侧。这些测点布置在垂直于飞行方向的直线上，两点之间的距离为 3 m。为了保证足够的视野，热红外成像仪固定在标记点前方，距离标记点 180 m。考虑到飞机的高速度，使用 Research IR 软件捕获帧频为 30 Hz 的图像序列，将可见光照相机放置在热红外成像仪附近以便于操作。

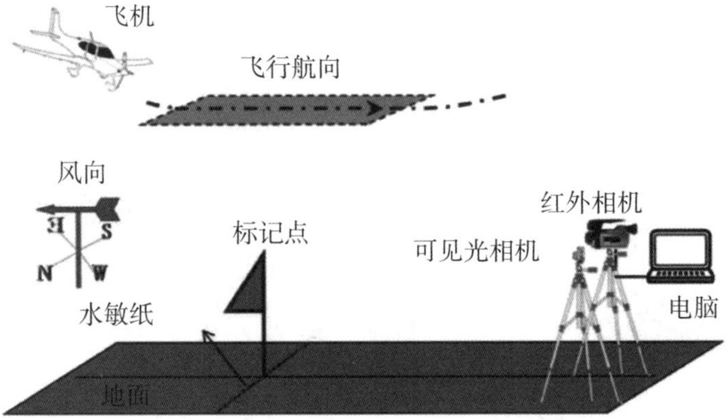

图 7　航空施药过程示意

尽管航空施药实验在野外进行，但考虑到飞机的高航速和天气条件，实验场中的背景温度在较短的施药过程中可被视为常数。施药雾滴的可见光图像和对应于红外热图像的灰度图像代表喷雾的浓度。图 8 表示在不同时刻进行空中喷雾之前和之后实验区域中的热辐射差异的灰度图像。飞行员调整飞机的飞行位置并在 0 ms 时刻开始释放雾滴。在 0 ms 至 2988 ms 的灰度图像中可以清楚地看到液滴的沉积过程：扩散到大气中的液滴浓度低于喷嘴附近的液滴浓度，大气中的灰色比喷嘴附近的灰色浅。因此，由大气中的液滴浓度引起的红外热辐射的差异小于喷嘴附近的红外热辐射的差异，液滴动态沉积过程的可视化为评估喷嘴性能和天气条件对喷雾质量的影响提供了可能。

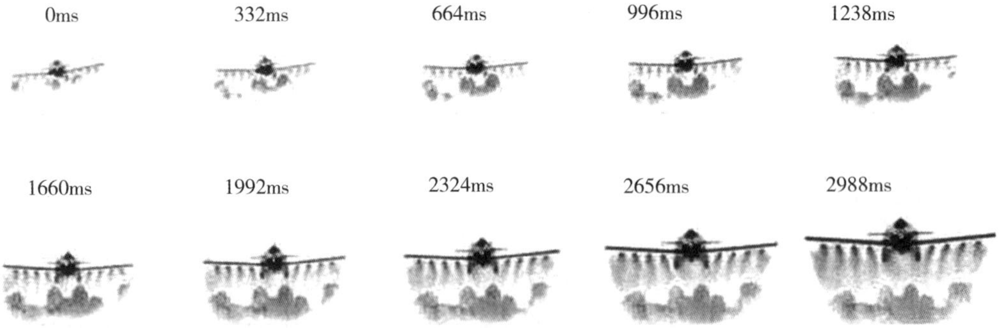

图 8　不同时刻实验区域热辐射灰度

如图 9 所示，为了能在所使用的方法和水敏纸方法之间进行视觉比较，我们提取了在 0 ms 到 2988 ms 的水敏纸附近具有相同区域的灰度图像。在图 9（a）中，从 0 ms 到 1238 ms 的提取灰度图像的颜色不明显。这种现象可能是由于液滴没有沉积到该区域中，或者由液滴浓度引起的热辐射差异太小而不能被检测到。然而，1660 ms 和 2988 ms 的灰色图像中明显的颜色变化表明液滴沉积在该区域并开始积聚。在图 9（b）中，累积了每个提取的灰度图像中包括的所有像素的灰度值。累积灰度值在 664 ms 之前几乎保持不变。结果表明，在 664 ms 之前液滴没有落入该区域；累积灰度值从 664 ms 到 996 ms 的小幅下降意味着液滴落入该区域并开始累积；在 996 ms 之后，由于该区域中液滴的逐渐沉积，累积的灰度值显著降低。

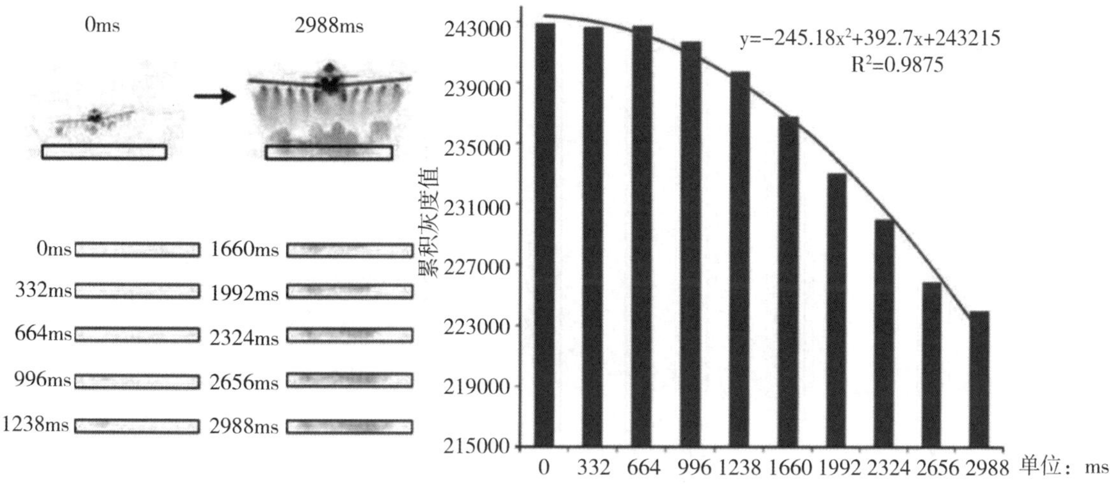

（a）不同时刻的红外热成像灰度图　　（b）累积灰度值随时间变化趋势

图9　不同时刻的红外热成像灰度

研究表明，基于红外热成像方法并结合图像处理算法能够检测喷雾过程中的微小热差异，从而监测雾滴空中飘移。本研究验证了红外热成像方法结合红外图像处理算法可用于监测航空喷雾应用中的农药飘移问题。该方法具有快速、非接触、实时测量的优点。

4.2　利用地面雾滴沉积传感器网络评估航空施药效果

针对实时测量雾滴沉积分布规律的需求，基于变介电常数电容器原理和传感器网络技术，本研究团队设计了航空施药雾滴地面沉积实时监测系统，并研究了该系统用于雾滴地面沉积量测量的实用性。相比于传统施药过程中的雾滴沉积分布特性检测，该方法具有可扩展性、实时性，可快速回传反馈，可记录雾滴蒸发等优势。[36]

该地面雾滴沉积传感器基于 Zigbee 通信技术设计，由传感器节点、协调器和数据分析计算机三部分组成。实时监测系统设备部署并启动完成后，根据网络规模和协调器与传感器节点的距离，传感器节点与协调器自组织形成 Zigbee 星形或树形拓扑网络。网内所有传感器节点数据都发送给协调器，协调器通过 USB 通信接口将数据发送给计算机，同时，计算机也可以经协调器将配置信息和指令信息发送给传感器节点。

2013 年 9 月 22 日，该系统在黑龙江省佳木斯某农用航空机场进行了实地应用测试。喷施液体选用电导率 487 μS/cm 自来水。地面共设置 14 个样点，在飞机航线左右各放置 7 个样点，采样间距 3 m。传感器节点如图 10 所示安装于支架上，水敏纸与传感器节点间隔 20 cm 平行放置。实验重复进行了 4 次，得到 56 个有效样点数据。

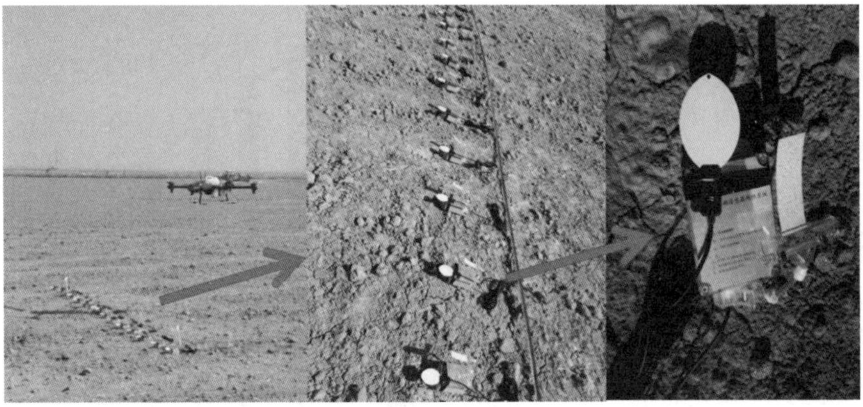

图 10　传感器样点实际布置

利用测量系统和文献[37]所述的水敏纸图像分析工具分别对 56 个样点的雾滴沉积量进行计算，得到如图 11 所示的两条数值分布曲线，可见两种方法对雾滴沉积量分布趋势测量结果的一致性较好。利用该系统测量得到的数据与基于变异系数分析方法得到的施药有效幅宽及雾滴沉积均一性数据相一致。需要指出的是，水敏纸图像分析方法的检测精度与雾滴粒径相关性较大，对于粒径小于 50 μm 的雾滴，其最大测量误差可达 34%；而对于粒径大于 1000 μm 的雾滴，其测量误差则仅有 12%。

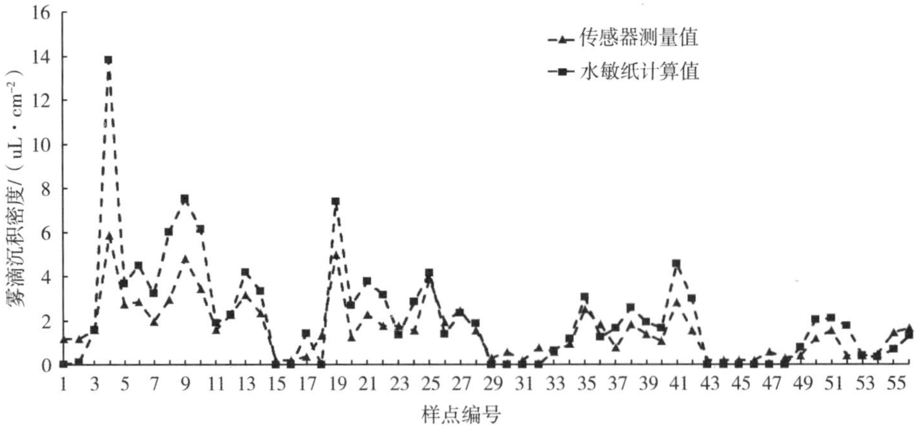

图 11　传感器与水敏纸测量雾滴沉积量对比

以水敏纸图像处理方法获得的数据为参照，从图 11 中可以看出，在测量系统对雾滴沉积量的量化测量值中，有 11 个点的测量相对误差小于 10%，占总样点数的 20%；有 8 个点的测量误差大于 50%，占总样点数的 14%；绝大多数点的测量相对误差在 10%～50%。从测量偏差来看，第 1 次数据测量偏差较大，第 4 次测量偏差较小。

结合水敏纸样点数据，分析得出测量偏差主要来自两个影响因素：一是雾滴沉积量的非均一性影响。受飞机飞行及气象影响，在飞机与地面间形成多变复杂的气流，气流使沉积量在较小距离范围内可能产生较大变化。现场的水敏纸数据表明，即使距离相近（小于 10 cm）的两个样点沉积量都会有较大差异。第二，雾滴的相互粘连也是重要的影响因素。当雾滴出现相互粘连时，可能导致传感器测量误差增大。

4.3 基于遥感影像评估航空施药效率

2013 年 7 月 6 日，团队在黑龙江省建三江农场进行航空施药实验，针对水稻冠层施药前后不同频率谱段变化特性开展研究。我们利用水敏纸对总共 45 个样地进行取样，并将其结果与光谱图像进行对照。在 3 个实验区域中，分别通过 ASD2500 光谱仪采集冠层光谱，通过成像光谱仪拍摄冠层高光谱图像，以及利用 Dualex 4 获取荧光参数和叶绿素含量（见图 12）。

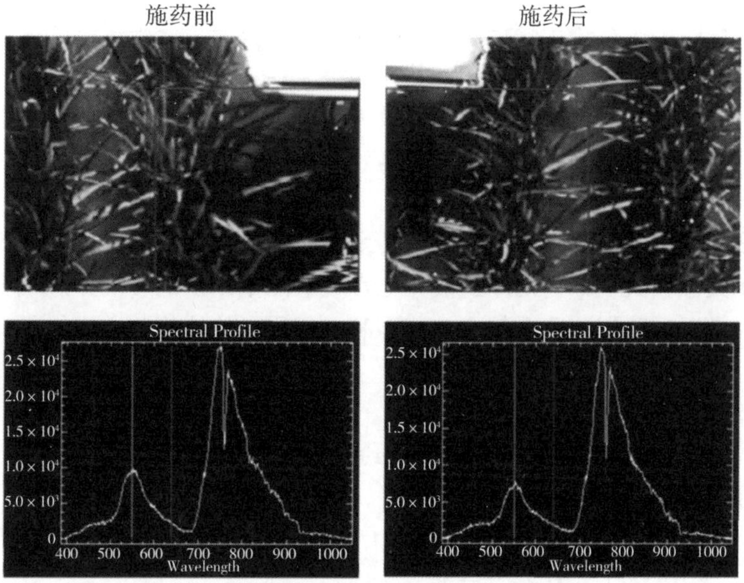

图 12　喷雾前后水稻冠层光谱成像结果对比

通过对喷雾前后作物光谱特性进行比对，我们发现，农药仅对水稻冠层产生了较少影响。这可能是由于这两个观察结果是及时进行的，并且作物没有受到病虫害胁迫。根据图 12 所示的光谱成像结果，在喷雾之前和之后，光谱各谱段能量没有显著变化。

通过对基于遥感影像评估航空施药效率进行探索研究，我们发现，基于三脚架测量的成像光谱仪视野狭窄，仅能覆盖非常小的作物区域，两个光谱测量的间隔很短。因此，农药喷雾的光谱响应可能不明显，延迟测量时间可能更合理。对于选择野外调查点，选择一些明显感染的区域和参考区域进行光谱测量是很重要的。

5　总结

针对航空精准施药实现及质量评价这一问题，我们进行了从雾滴飘移预测到施药作业导航、地面沉积检测的体系化研究。航空施药导航控制及变量施药系统通过整合导航作业系统、变量施药控制系统、可控转速雾化器、环境风速传感器、雾滴飘移预测模型等，提升了机载精准施药系统的智能化水平。我们基于 CFD 技术，构建了航空施药雾滴飘移和沉积预测模型，模型能够提供雾滴粒径、侧风风速、飞机尾涡、雾化器位置、温度和相对湿度等因素对于雾滴飘移沉积的影响规律。本项目组采用了红外热像检测雾滴飘移分布和喷洒范围的方法，开发了航空施药地面沉积传感器网络系统，用于实时测量航空施药雾滴沉积、蒸发动态信息，探索了遥感光谱图像在航空施药雾滴沉积质量评价中的应用可行性。

未来，我们将重点围绕以下几个问题进一步开展研究。

一是变量施药控制系统及导航作业系统的控制精度研究。由于变量施药涉及采集信号、控制决策、机械响应、药泵和管道流体相应、超调修正等多个环节，特别是药泵和管道流体在控制系统是大时滞控制对象，而飞机飞行速度较快，变量施药控制系统响应时间是制约飞机机载航空变量施药控制精度的关键问题。今后，我们将通过综合惯导数据、先进控制算法等进一步优化控制系统响应时间。

二是航空施药雾滴飘移和沉积预测模型优化研究。目前的研究仍基于二维 CFD 模拟，需要数值求解 N-S 方程，其计算消耗较大，难以进行雾滴飘移的实时预测。未来需进一步优化算法和模拟方案，克服模拟精度和模拟速度的矛盾。

三是航空施药雾滴沉积质量测量精度研究。从当前测试实验结果的对比来看，基于电子传感器技术的施药雾滴沉积质量测量技术更具有实用性，但是该系统测量精度与药液介电常数相关性较大，如何消除药液介电常数对测量结果的影响，进一步提升测量结果的精准性，是以后我们团队在航空施药雾滴沉积质量快速检测方面的研究重点。

参考文献

[1] 傅泽田，祁力钧，王俊红. 精准施药技术研究进展与对策 [J]. 农业机械学报，2007，38 (1)：189 -192.

[2] 张东彦，兰玉彬，陈立平，等. 中国农业航空施药技术研究进展与展望 [J]. 农业机械学报，2014，45 (10)：53 -59.

[3] 朱传银，王秉玺. 航空喷雾植保技术的发展与探讨 [J]. 植物保护，2014 (5)：1 -7.

[4] 周志艳，明锐，臧禹，等. 中国农业航空发展现状及对策建议 [J]. 农业工程学报，2017，33 (20)：1 -13.

[5] 周志艳，臧英，罗锡文，等. 中国农业航空植保产业技术创新发展战略 [J]. 农业技术与装备，2014，29 (5)：1 -10.

[6] 罗锡文. 对加快发展我国农业航空技术的思考 [J]. 农业技术与装备，2014 (5)：7 -15.

[7] 薛新宇，兰玉彬. 美国农业航空技术现状和发展趋势分析 [J]. 农业机械学报，2013，44 (5)：194 -201.

[8] ZHANG M, ZHANG R, XU G, et al. Design and development of a navigation system for agricultural aerial spraying [C]. Edinburgh: ECPA 2017 - the 11th European conference on precision agriculture, 2017.

[9] 张瑞瑞，李杨，伊铜川，等. 有人直升机变量施药控制系统的设计与实验 [J]. 农机化研究，2017，39 (10)：124 -127.

[10] CRABBE R, McCOOEYE M, MICKLE R. The influence of atmospheric stability

on wind drift from ultra-low-volume aerial forest spray applications [J]. Journal of applied meteorology, 1994, 33: 500-507.

[11] TESKE M, BIRD S, ESTERLY D, et al. AgDRIFT: a model for estimating nearfield spray drift from aerial applications [J]. Enviromental toxicology and chemistry, 2002, 21 (3): 659-671.

[12] YATES W, AKESSON N, COWDEN R. Criteria for minimizing drift residues on crops downwind from aerial applications [J]. Transaction of the ASAE, 1974, 17: 627-632.

[13] YATES W, AKESSON N, COUTTS H. Evaluation of drift residues from aerial applications [J]. Transaction of the ASAE, 1966, 9: 389-393.

[14] YATES W, AKESSON N, COUTTS H. Drift hazards related to ultra-low-volume and dilute sprays by agricultural aircraft [J]. Transaction of the ASAE, 1967, 10: 628-632.

[15] WARE G, CAHILL W, ESTESEN B. Pesticide drift: aerial applications comparing conventional flooding vs. raindrop nozzles [J]. Journal of economic entomology, 1974, 68: 329-330.

[16] BIRD S, ESTERLY D, PERRY S. Off-target deposition of pesticides from agricultural aerial spray applications [J]. Journal of environmental quality, 1996, 25: 1095-1104.

[17] ZHANG B, TANG Q, CHEN L, et al. Numerical simulation of spray drift and deposition from a crop spraying aircraft using a CFD approach [J]. Biosystems engineering, 2018, 166: 184-199.

[18] ZHANG B, TANG Q, CHEN L, et al. Numerical simulation of wake vortices of crop spraying aircraft close to the ground [J]. Biosystems engineering, 2016, 145: 52-64.

[19] 张慧春, 郑加强, 周宏平, 等. 转笼式生物农药雾化喷头的性能实验 [J]. 农业工程学报, 2013, 29 (4): 63-70.

[20] 吕晓兰, 傅锡敏, 吴萍, 等. 喷雾技术参数对雾滴沉积分布影响实验 [J]. 农业机械学报, 2011, 42 (6): 70-75.

[21] FRITZ B, HOFFMANN W, BAGLEY W. Effects of spray mixtures on droplet size under aerial application conditions and implications on drift [J]. Applied engineering in agriculture, 2010, 26 (1): 21-29.

[22] KIRK I, HOFFMANN W, FRITZ B. Aerial application methods for increasing spray deposition on wheat heads [J]. Applied engineering in agriculture, 2006, 23 (6): 357-364.

[23] BIRD S, ESTERLY D, PERRY S. Off-target deposition of pesticides from agricultural aerial spray applications [J]. Journal of environmental quality, 1995, 25 (5): 1095-1104.

[24] ZHANG H, LAN Y, LACEY R, et al. Ground-based spectral reflectance measurements for evaluating the efficacy of aerially-applied glyphosate treatments [J]. Biosystems engineering, 2010, 107 (1): 10-15.

[25] HUANG Y, HOFFMANN W, FRITZ B, et al. Development of an unmanned aerial vehicle-based spray system for highly accurate site-specific application [C]. Rhode Island: 2008 ASABE annual international meeting, 2008.

[26] HUANG Y, HOFFMANN W, LAN Y, et al. Development of a spray system for an unmanned aerial vehicle platform [J]. Applied engineering in agriculture, 2009, 25 (6): 803 - 809.

[27] FISHER R, MENZIES D, HERNE D. Parameters of dicofol spray deposit in relation to mortality of European red mite [J]. Journal of economic entomology, 1974, 67 (1): 124 - 126.

[28] 张京, 何雄奎, 宋坚利, 等. 无人驾驶直升机航空喷雾参数对雾滴沉积的影响 [J]. 农业机械学报, 2012, 43 (12): 94 - 96.

[29] 秦维彩, 薛新宇, 周立新, 等. 无人直升机喷雾参数对玉米冠层雾滴沉积分布的影响 [J]. 农业工程学报, 2014, 30 (5): 50 - 56.

[30] FRANZ E, BOUSE L, CARLTON J. Aerial spray deposit relations with plant canopy and weather parameters [J]. Transaction of the ASAE, 1998, 41 (4): 959 - 966.

[31] 茹煜, 贾志成, 范庆妮. 无人直升机远程控制喷雾系统 [J]. 农业机械学报, 2012, 43 (6): 47 - 52.

[32] 吕晓兰, 傅锡敏, 宋坚利, 等. 喷雾技术参数对雾滴飘移特性的影响 [J]. 农业机械学报, 2011, 42 (01): 59 - 63.

[33] HOFFMANN W, KIRK I. Spray deposition and drift from two "medium" nozzles [J]. Transactions of the ASAE, 2005, 48 (1): 5 - 11.

[34] GULER H, ZHU H, OZKAN H, et al. Wind tunnel evaluation of drift reduction potential and spray characteristics with drift retardants at high operating pressure [J]. Journal of ASTM international, 2006, 3 (5): 1 - 9.

[35] JIAO L, DONG D, FENG H, et al. Monitoring spray drift in aerial spray application based on infrared thermal imaging technology [J]. Computers & electronics in agriculture, 2016, 121 (C): 135 - 140.

[36] 张瑞瑞, 陈立平, 兰玉彬, 等. 航空施药中雾滴沉积传感器系统设计与实验 [J]. 农业机械学报, 2014, 45 (8): 123 - 127.

[37] WOLF R, GARDISSER D, LOUGHIN T. Comparisons of drift reducing/deposition aid tank mixes for fixed wing aerial applications [J]. Journal of ASTM international, 2006, 2 (8): 14.

作者简介

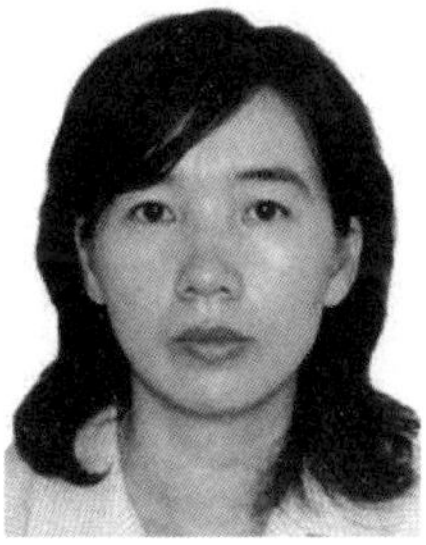

陈立平，国家农业智能装备工程技术研究中心主任，研究员，博士生导师。国际农业工程学会（CIGR）农业航空应用委员会成员，美国农业和生物工程师学会（ASABE）药液喷洒应用委员会成员，中国农业工程学会农业航空分会副主任委员，中国农业机械学会农业航空分会副主任委员。《农业工程学报》《农业机械学报》、*International Journal of Agricultural and Biological Engineering* 等学术期刊编委；入选国家“万人计划”领军人才、国家百千万人才工程、科技部中青年科技创新人才计划，国务院政府特殊津贴专家。其作为负责人建立的中美农业航空联合技术中心是我国第一个国家农业航空应用技术国际联合研究中心。发表学术论文 91 篇（SCI/EI 收录 56 篇），主编专著 2 部，参编专著 6 部，获 26 项发明专利、27 项实用新型专利和 39 项软件著作权。主要从事农业智能装备技术研究工作。

张瑞瑞，博士，副研究员，主要从事农业航空精准施药技术与装备系统研究。以第一作者/通信作者发表农业航空应用技术相关学术论文 32 篇（SCI/EI 26 篇）；获国家发明专利授权 24 项、国家软件著作权登记 41 项。研发完成的航空施药雾滴沉积监测与分析系统、航空施药作业监管与自动计量系统、飞机机载施药作业导航与管理系统、飞机机载变量施药精准控制器等在行业内得到推广应用和产业化。

唐青，博士，副研究员，主要从事固定翼有人飞机、旋翼无人机的航空施药风场及雾滴运动的模拟与试验研究。发表 SCI/EI 论文 7 篇。

徐刚，博士，主要从事农业航空喷雾质量检测技术及导航应用软件系统研究。发表 SCI/EI 论文 12 篇。

Robotics in Precision Agriculture

Dimitris Zermas① David Mulla②

1 Introduction

Precision agriculture (PA) is a management practice applied at the right rate, right time and the right place. It replaces uniform management of farms with customized sub-region management. PA is helping to better manage nutrients, irrigation, pests and weeds, tillage and seeding operations. Benefits of PA include increased profitability, improved efficiency of inputs, improved yield and quality of crop, reduced risk, and protection of the environment.[2]

Nitrogen (N) deficiency in corn results in a 20% reduction of the yearly yield in Midwestern corn fields. To avoid N stresses entirely, farmers currently apply excessive amounts of N fertilizer to corn, leading to pollution of local and regional water sources. Approaches are needed to identify N deficiency in corn during early growth stages, followed by variable rate N applications. Several spectral indexes that use combinations of different reflectance bands have been used to identify the existence of specific deficiencies. Bhatti, et al.[2] utilized satellite imagery to assess the nutrient state of bare soil; this was the first documented attempt of integrating in PA applications. In a later study, Blackmer, et al.[3,4] discovered that N deficiencies become apparent when computing the ratio of light reflectance between 550 nm and 600 nm divided by the reflectance between 800 nm and 900 nm and developed an index to describe their severity based on the chlorophyll absorption. A similar discovery considering chlorophyll absorption for wavelengths in the visible spectrum (430 nm and 660 nm) was presented by Pinter, et al.[5]. In addition, to remove interference from soil, segmentation methods for the isolation of crop vegetation from the background soil were developed based on panchromatic images and row planting geometry.[6]

Modern approaches that employ unmanned aerial vehicles (UAVs) flying at upper limits of U. S. altitude restrictions produce stitched field images of crop spectral indices based on visible

① Dimitris Zermas, Dept. Computer Science and Engineering.
② David Mulla, Dept Soil, Water and Climate University of Minnesota.

and near infrared (NIR) reflectance. Fixed-wing and rotary UAVs have both been used for low cost surveillance of farms. Although these approaches are able to locate problematic areas, existing methods are unable to differentiate N stress from potassium, sulfur, or water stress in the crop. The methodology introduced in this work utilizes high resolution imagery collected using low flying UAVs to specifically detect N deficiencies in corn plants and assess its severity at an early growth stage.

2 Methods

Nitrogen deficiency tends to exhibit unique V-shaped yellow signatures on corn leaves (see Fig. 1), and crop scouts use those signs to identify N stress. Our goal is to translate the expertise of agronomists to machine language, to enhance and speed up the detection of N stresses. We are utilizing the latest advances in computer vision (CV) and mature machine learning (ML) approaches to address this problem in a real world environment.

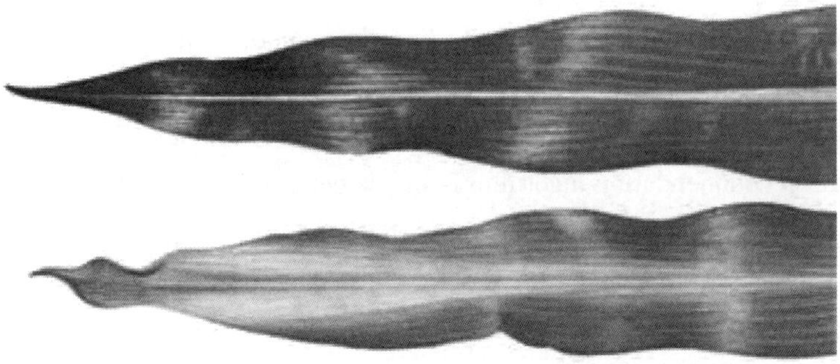

Figure 1 Comparison between corn leaves exhibiting N stress (bottom) and no N stress (top)

The collection of high resolution panchromatic imagery in the visible spectrum took place over the course of three cultivation periods (2014 – 2016) with data capturing a combination of corn growth stages varying from V5 to V12. Two maize fields in the state of Minnesota were selected for testing our methodology. Two very different soil types were considered, namely, sandy and clay loam. The soil type greatly affected the number of visible deficient leaves during the early stages (V5 and V8).

Data were collected using two different platforms: a MikroKopter Okto XL equipped with a Nikon D7200 RGB sensor and a DJI Matrice 100 with a Zenmuse X3 RGB high resolution camera. A total of 108 images were collected at growth stages V5, V8, and V12 from plots that were treated with 6 different N fertilizer rates (0 kg/ha, 66 kg/ha, 132 kg/ha, 196 kg/ha, 264 kg/ha, and 330 kg/ha). Out of these, the images that showed solely healthy plants were discarded as we are interested in identifying N deficiencies only for stressed plants. The resolution of the images varies within a range of 0.2 to 0.47 cm/pixel.

An accurate method was developed for the detection of N deficiency in corn plants during the early stages of their development (V6 – V12). We used a UAV to collect high resolution visible (RGB) images during a low altitude flight (< 15m). The flight was semi-automated with waypoints provided beforehand and the camera is triggered once the UAV reaches a waypoint.

The methodology consists of two parts: ① the detection of rectangles of interest (ROI) in each image, and ② the prediction of N deficiency inside the ROI. The first part results in locating the areas in the form of rectangular regions exhibiting potential N deficiency, much as a digital camera uses facial recognition to place a rectangle around potential human faces in a view screen. The second part acts as a filter on the output from the first module, further refining the decision of correctly identifying areas of leaves showing N deficiency with a high degree of confidence.

2.1 Recommendation Scheme

The first step towards identifying N-deficient leaves in images is to recommend image regions that hold significant information regarding the general state of health of the individual leaves. This is an important step in the process pipeline, because it limits the computations to only small image areas, thus increasing performance and reducing computational time. The concept behind a recommendation algorithm is the selection of meaningful subsets of pixels from a given image and this work identifies such subsets based on homogeneity of their color spaces. This algorithm includes the following three steps:

- a hierarchical unsupervised clustering that produces two major clusters,
- a semi-supervised classification that utilizes a Support Vector Machine model, and
- several low level morphological operations which assist in the refinement of this module's output.

Pixels are clustered based on color and partitioned into 3 classes. The first class consists of green pixels, the second of yellow pixels (associated with potential N deficiencies), and the last of pixels that correspond to the soil. Using an unsupervised clustering algorithm tends to naturally separate green pixels from the other dominant colors, namely yellow and brown. This algorithm initially aims to segment only the green pixels of the image by employing a hierarchical K-means algorithm on the combined space of two color spaces: RGB and L * a * b[7].

The clustering module involves two steps, first a clustering step that breaks down the image into multiple groups of similarly colored pixels, and second a 2-group clustering that separates the green parts of the plants from the rest of the image. For the selection of the appropriate clustering algorithm, we considered attributes such as the complexity with respect to the number of queries and the utilization of physical memory during the computation process. The unsupervised K-means algorithm met the requirements for the task, while providing satisfactory performance.

When dealing with NIR spectra, the concept of the soil line[8] is widely used and provides an accurate index for the identification of the soil pixels. On the other hand, in the visible spectrum the automated distinction between yellow pixels and pixels belonging to the soil proves to be particularly challenging. Once green pixels are separated from the rest, a semi-supervised classification scheme partitions the non-green pixels into yellow and soil clusters.

The process developed asks the user to draw a single rectangle around pixels that represent soil and another rectangle around yellow pixels which represent leaves with N deficiency. Subsequently, a Support Vector Machine (SVM) classification model with a linear kernel is trained based on these selections and is used for the classification between yellow and soil in the rest of the images.

Finally, connected regions of yellow pixels are smoothed by morphological operators and bounding boxes are assigned to them in an effort to highlight informative candidate regions. Specifically, parts of the plant that are one pixel apart are connected in order to remove discontinuities while morphological opening and closing are used as hole filling techniques for the removal of empty patches inside the hull of the leaves. The smooth and symmetrical objects that result from these morphological operations guarantee high performance of the feature extraction step described in the next section. An additional morphological step removes small groups of pixels based on a threshold that considers their size. The threshold is manually selected through a trial and error process and can fluctuate depending on the resolution of the initial image. The surviving leaves in this recommendation scheme for potential N stress (see Fig. 2) are the candidates for the second module, which further defines the regions showing potential N stress.

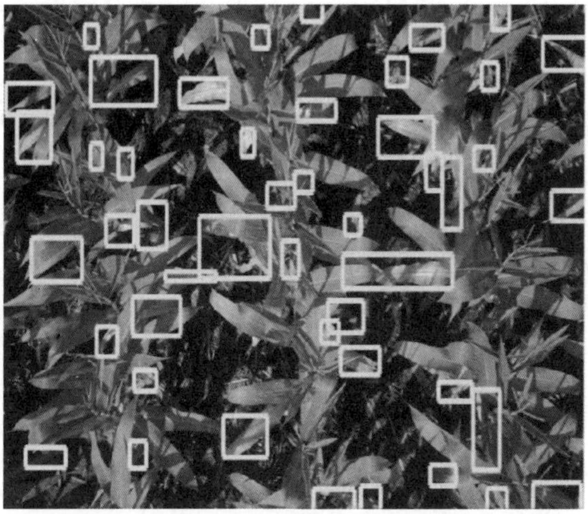

Figure 2 Recommendation scheme for leaves exhibiting potential N stress

2.2 N Deficiency Assessment

Prior to this step, the recommendation scheme described in part 2.1 above selects candidate image regions that potentially exhibit nitrogen deficiencies. Due to the structure of the recommendation scheme an additional inference step is used to distinguish deficient regions from regions depicting tassels, healthy leaves and stressed leaves that cannot be assessed due to the advanced state of their condition (e. g. completely yellow leaves near the soil). For this task we developed a supervised learning scheme that provides a label to the candidate regions. Positive labels correspond to regions containing nitrogen deficient leaves while negative labels correspond to non-deficient leaves and other plant parts. In this step, an assessment regarding the deficiency of a candidate region is performed. The inputs consist of rectangular regions suggested by the recommendation algorithm, and depicts potentially affected parts of the image. Among the selected candidates, the leaves that exhibit N deficiency are separated from the rest (e. g. tassels or yellow leaves near the soil).

3 Results

Following a hierarchical K-means clustering, the green cluster was identified with an average accuracy of 96.3%. Specifying a low number of clusters in the multi-class clustering phase speeds up processing time, while maintaining high accuracy for days with different solar illumination. Based on those results, seven classes were selected for the multi-cluster K-means approach.

Segmenting the green parts in a single color space was not accurate for all images. This is especially true in cases with a few N deficiencies, where a significant variance in the representation of the green color is presented. The accuracy of the segmentation increases when combining clustering results for two color spaces. This method achieves robust results in the segmentation of green pixels for all the subject images. Next, the distinction between yellow areas and soil is performed by an SVM classifier. The SVM model is sensitive enough to identify individual yellow pixels, and the thresholds for morphological operations can be customized by the user to match the desired granularity of the results.

Elimination of ROI that did not exhibit N deficiency resulted in almost perfect classification, with accuracies of 96.1% ~ 99.2%. These results validate the hypothesis that the distinct yellow color and the V-shape of N deficiency in maize are captured by the two descriptors.

4 Discussion

This research shows that UAVs equipped with a high resolution digital camera are capable of detecting maize leaves that exhibit N deficiency at early growth stages (V6 – V8). In practice, the UAVs could fly at an altitude of 100 m to detect areas of a field that appear to have canopy scale yellowing, and then fly low (altitude of 20 m or less), to determine whether or not the yellowing is caused by N deficiency. If it is N deficiency, the extent and severity of crop stress can be determined, allowing the proper rate of in-season N fertilizer recommendation to be made.

Previous attempts to detect N deficiency with UAVs have relied primarily on Normalized Difference Vegetation Index (NDVI) values obtained with multi-spectral cameras flown at altitudes of 100 m or higher. NDVI imagery of this type is at the canopy scale, and is unsegmented so that interference from bare soil often occurs. In addition, low NDVI values, which indicate crop stress, could arise not only from N stress, but also from stresses due to potassium and sulfur deficiency, as well as water logging of crops[9]. In contrast, our method specifically detects N deficiency in maize.

Over 3,000 UAVs manufacturing companies are in China. Many of these UAVs are capable of carrying large payloads, and are used to spread a uniform rate of fertilizer on agricultural fields in China. However, it is rare in China to find UAVs that spread fertilizer and also are equipped with cameras to detect N deficiency. In principle, combining our method for detecting N deficiency in corn could be implemented on UAVs in China that spread N fertilizer. This combination would allow in-season topdress applications of N fertilizer to be used as an alternative to uniform spreading of N fertilizer. This could lead to improvements in crop yield, greater N use efficiency and reductions in environmental N pollution as a result of applying the right rate of N fertilizer at the right time to better synchronize N fertilizer application with crop N uptake patterns.

5 Conclusions

In this work, a complete stand-alone computer vision tool for the detection of N deficiency in the early growth stages of corn plants has been developed, and validated on an extensive dataset with real world images at a resolution of less than 0.47 cm/pixel.

Two distinct modules were developed as part of this tool, one to detect color based on ROI inside a given image, and another to assess the existence of N deficiency inside that ROI. The first module successfully segmented the color space of a corn field image into three dominant classes representing the green parts of the plants, the yellow areas which may contain deficiency, and the soil, under a variety of image resolutions and illuminations. N deficiencies in images were detected with an accuracy of 96.1% ~ 99.2%.

References

[1] MULLA D J. Twenty five years of remote sensing in precision agriculture: key advances and remaining knowledge gaps [J]. Biosystems engineering, 2013, 114: 358 -371.

[2] BHATTI A U, MULLA D J, FRAZIER B E. Estimation of soil properties and wheat yields on complex eroded hills using geostatistics and thematic mapper images [J]. Remote sensing of environment, 1991, 37: 181 -191.

[3] BLACKMER T, SCHEPERS J. Use of a chlorophyll meter to monitor nitrogen status and schedule fertigation for corn [J]. Journal of production agriculture, 1995, 8: 56 -60.

[4] BLACKMER T, SCHEPERS J, VARVEL G, et al. Nitrogen deficiency detection

using reflected shortwave radiation from irrigated corn canopies [J]. Agronomy journal, 1996, 88: 1-5.

[5] PINTER JR P J, HATFIELD J L, SCHEPERS J S, et al. Remote sensing for crop management [J]. Photogrammetric Engineering & Remote Sensing, 2003, 69: 647-664.

[6] ONYANGO C M, MARCHANT J A. Segmentation of row crop plants from weeds using colour and morphology [J]. Computers and electronics in agriculture, 2003, 39: 141-155.

[7] FAITHPRAISE F, BIRCH P, YOUNG R, et al. Automatic plant pest detection and recognition using k-means clustering algorithm and correspondence filters [J]. International journal of advanced biotechnology and research, 2003, 4: 189-199.

[8] BARET F, JACQUEMOUD S, HANOCQ J F. The soil line concept in remote sensing [J]. Remote sensing reviews, 1993, 7: 65-82.

[9] CLAY D E, KIM K I, CHANG J, et al. Characterizing water and nitrogen stress in corn using remote sensing [J]. Agronomy journal, 2006, 98: 579-587.

作者简介

Dimitris Zermas received his Diploma of Engineering in Electrical Engineering from the University of Patras, Greece in 2012. He is a Ph. D. candidate at the department of Computer Science and Engineering at the University of Minnesota. His research interests include geometric vision, image processing, machine learning and robotics with applications in precision agriculture. Currently, he is a doctoral dissertation fellow at the University of Minnesota.

David Mulla received a Ph. D. degree in Agronomy from Purdue University with emphasis in soil chemistry and soil physics (1983). He is currently Professor and Larson Chair for Soil & Water Resources in the Department of Soil, Water, and Climate at the University of Minnesota, and Director of the Precision Agriculture Center. Dr. Mulla is an internationally recognized researcher and scholar. His research has taken him to over 20 countries. Dr. Mulla is widely recognized as a pioneer in precision agriculture, based on seminal research on the topics of management zones, spatial statistics, remote sensing and variable rate fertilizer application. He has been an associate editor of *J. Prec. Ag* since 1997; was elected a Fellow in Soil Science Society of America (SSSA) and Agronomy Society of America (ASA) in 1997 and 1999, respectively; was awarded the P. C. Robert Senior Precision Agriculture Research Award in 2012 by the International Society for Precision Agriculture; and was awarded the Soil Science Applied Research Award in 2013 by the SSSA.